BIOCHEMICAL PHYSICS: FRONTAL RESEARCH

BIOCHEMICAL PHYSICS: FRONTAL RESEARCH

SERGEI D. VARFOLOMEEV
ELENA B. BURLAKOVA
ANATOLII A. POPOV
GENNADY E. ZAIKOV
EDITORS

Nova Science Publishers, Inc.
New York

Library of Congress Cataloging-in-Publication Data
Available upon request.

ISBN 13 978-1-60021-425-7
ISBN 10 1-60021-425-8

Published by Nova Science Publishers, Inc. ✣New York

CONTENTS

PREFACE

This book presents recent research in chemical and biochemical physics. Chemical physics addresses a large range of problems. An effective chemical physicist is a "jack-of-all-trades," able to apply the principles and techniques of the field to everything from high-tech materials to biology. Just as the fields of chemistry and physics have expanded, so have chemical physics subject areas, which include polymers, materials, surfaces/interfaces, and biological macromolecules, along with the traditional small molecule and condensed phase systems. Biochemical physics is a science that joins the three natural sciences biology, chemistry and physics into one comprehensive study. N.M. Emanuel pioneered this science over fifty years ago. This book presents papers written by Emanuel's students that reveal recent developments in this interesting field.

In Chapter 1, parameters of inhibited oxidation of isotactic propylene containing antioxidant 2,2'-methylene-bis(4-methyl-6-tert.butylphenol) has been studied in the temperature range 140 – 210° C. Unlike polyethylene, the dependencies of inhibited oxidation parameters: of critical antioxidant concentration and of rate constant of antioxidant consumption change with temperature do not obey Arrhenius law both below and above the melting temperature.

The vulcanization of 1,4-cis-polyisoprene was studied by differential scanning calorimetry (DSC) under isothermal and non-isothermal conditions. On the basis of thermal characteristics obtained in chapter 2, the kinetic parameters of cross-linking (the induction period, maximum rate, and effective energy of activation) were determined.

The investigation results of the simultaneous effect of electric current and tensile deformation on the oxidation processes and electroconductive polyisoprene composition (EPIC) properties are presented in chapter 3.The effects of the EPIC oxidation process under the conditions of the complex factor action consisting of the gaseous product emission and decrease of the oxidation rate at the final stage of the experiment were disclosed.It was found out that the change in the electric resistance with time of EPIC aging was characterized by three sections: decrease, constant value area and increase of the electric resistance. It was demonstrated that the values of the electric resistance and duration of the above sections depended on the deformation temperature and extent.

A regulatory peptide (RP) biologically active at ultralow doses (10^{-12} mg/ml) was isolated from the bovine eye lens and studied in chapter 4 with respect to amino acid composition and secondary structure. This RP was shown to have a protective effect on eye cells in experiments on chemically induced cataract in amphibian and mammalian lenses cultured *in vitro* and traumatic cataracts development *in vivo* in rats. Using an immunohistochemical method, the RP was localized in the intercellular space of the lens epithelium.

A protein extract from the rat liver was analyzed for the presence of regulatory proteins (RPs) displaying biological activity at ultra low doses. Four fractions containing such proteins were identified, and their physicochemical properties were studied in chapter 5. To assess the specific activity of RPs, an original in vitro model of roller organ culture of the newt liver was developed. Using this model, the acidic hepatic regulatory protein was shown to have a protective effect on the liver.

In chapter 6, a regulatory protein (RP) biologically active at ultra low doses was isolated from the bovine cornea and analyzed for physicochemical properties and activity in biological tests. The RP was localized in the intercellular space of corneal epithelium and endothelium in rats and newts, and its content was shown to increase with time in organ cultures of the cornea. This RP proved to stimulate cell proliferation and migration in the corneal epithelium and to aid in the maintenance of corneal tissue structure *in vitro.*

Three fractions containing regulatory proteins (RPs) biologically active at ultra low doses weakly acidic, acidic, and basic were isolated from bovine blood serum. The results of electrophoresis in PAAG showed that a low-molecular-weight protein ($R_f = 0.9 \pm 0.05$) was the main component of all these fractions. According to circular dichroism spectra, the secondary structure of RPs was characterized by the prevalence of β-sheet regions and unordered structures described in terms of random coil, with the content of α-helices being low. Tolerance of RPs to the impact of various physicochemical factors was noted in the course of the study. An analysis by the method of atomic power microscopy showed that RPs in solutions of different concentrations were in the form of spherical nanoparticles. The results in chapter 7 provide evidence that RPs from bovine serum are largely analogous to each other in their physicochemical properties.

Regulatory proteins (RPs) detected in fractions of bovine blood serum-acidic (pI < 3.0), weakly acidic (pI 4.5–5.1), and basic (pH 8.5–9.5) - were analyzed for specific biological activity *in vivo* and *in vitro*. In addition to chapter 8, immunohistochemical localization of the weakly acidic RP in tissues was performed. This RP was detected in the intercellular space of the newt liver, which was indicative of its synthesis in this organ. All blood serum RPs at ultra low doses exhibited a protective effect on *in vitro* cultures of the newt skin. The weakly acidic RP (pI 4.5–5.1) proved to stimulate skin repair after injury, facilitating wound healing in mice.

The flammability characteristics of PP, PP nanocomposite based on layered silicate (Cloisite 20A), composition of PP with MH (50% wt.) and the composition of PP nanocomposite with 20 % wt. of MH were examined with a cone calorimeter at an external heat flux of 35 kW/m^2. The peaks heat release rate of PP/ MH (20% wt.) and PP–MPP-Cloisite 20A (5.6% wt.) with MH (20% wt.) decrease by a factor of six and four, respectively, in comparison with pristine PP. It means that relatively small incorporations of MH to PP nanocomposites can substantially reduce the important factor of combustibility, such as maximum rate of heat release. The main purpose of the present work in chapter 9, was the compare the flame retardancy of PP, PP/ MH (20% wt.) and PP–MPP-Cloisite 20A (5.6%

wt.) with MH (20% wt.) in order to obtain much better flame resistance using a small loading of MH.

In: Biochemical Physics Frontal Research
Editors: Sergei D. Varfolomeev et al., pp. 1-5
ISBN 1-60021-425-8

Chapter 1

VARIATION OF INHIBITED OXIDATION PARAMETERS IN THE RANGE OF PHASE TRANSITION

I. G. Kalinina*, K. Z. Gumargalieva* and Yu. A. Shlyapnikov

*Institute of Chemical Physics, Russian Academy of Sciences
**Institute of Biochemical Physics. Russian Academy of Sciences
119991, Moscow, Russia

ABSTRACT

Parameters of inhibited oxidation of isotactic propylene containing antioxidant 2,2'-methylene-bis(4-methyl-6-tert.butylphenol) has been studied in the temperature range 140 – 210° C. Unlike polyethylene, the dependencies of inhibited oxidation parameters: of critical antioxidant concentration and of rate constant of antioxidant consumption change with temperature do not obey Arrhenius law both below and above the melting temperature.

Polypropylene is sensitive to oxidation of atmospheric oxygen and needs stabilization against this process. As the most polymers, materials based on polypropylene are designed for long use, but investigations of the antioxidative stability of these polymers are usually performed at elevated temperatures, because the experiments at the temperatures of service may need many years. Thus to apply the data obtained at accelerated testing to the service conditions we must know how these data change in the range dividing the temperatures of testing and the temperatures of service.

In this study we investigated oxidation of isotactic polypropylene (IPP) prepared on metallocene catalysts in the range of 140° to 210° C and oxygen pressure 300 mm Hg. Antioxidant was 2,2'-methylene-bis(4-methyl-6-tert.butylphenol (MBP), often used as standard antioxidant.

Isotactic polypropylene is crystalline polymer. Crystalline polymer IPP does not completely crystallize at any temperature, though over the certain *melting temperature* the crystalline substance in it disappears completely. Some processes in the polymer proceed mainly or completely in amorphous phase, but the elements of the crystalline phase, even when they are small, are linking several polymer chains preventing their translational mobility and, in the same time, only slightly affect the vibrational one, Thus the intercrystalline amorphous substance in the polymer must resemble gel in which mobility of the polymer chains needed for formation of centers of the additive sorption is restricted. It is interesting to understand how these elements of crystalline phase affect the oxidation processes in polymer.

In this work wee studied dependencies of two partameters of inhibited oxidation: of the critical antioxidant concentration, below which antioxidant only slightly affects the oxidation rate, and of apparent rate constant of the antioxidant consumption on the temperature oxidation.

EXPERIMENTAL

IPP, $M_w = 690\ 000$, $M_w/\ M_n = 3{,}5$ and isotacticity 96% prepared by polymerization on metallocene catalyst [1] has been used. Antioxidant MBP has been purified by vacuum sublimation. Antioxidant was added to the polymer powder from alcohol solution with subsequent drying at room temperature. The samples of 0.050 g of IPP were oxidized in the vacuum equipment described in [2], volatile products of oxidation were absorbed by solid KOH.

To calculate the parameters of inhibited oxidation the formula (1) has been used [2, 3]

$$\tau = \tau_{o+} \frac{1}{k_{eff}} \ln \frac{[IH]_o}{[IH]_{cr}}, \qquad (1)$$

where τ - is induction period of oxidation of the sample containing antioxidant at concentration of $[IH]_o$, $[IH]_{cr}$ - the critical, i.e. the minimal effective concentration τ_o – induction period at this concentration (i.e. when $[IH]_o = [IH]_{cr}$), and k_{eff} - apparent rate constant of antioxidant consumption, describing the complex process initiated by direct oxidation of antioxidant [2].

According formula (1), the experimentally found dependence of induction period on initial antioxidant concentration must be straight line beginning from the horizontal line corresponding to induction period of non-inhibited polymer at $[IH]_o = [IH]_{cr}$ in the coordinates τ - $\lg[IH]_o$. The slope of this line is connected with the rate constant k_{eff} by the formula:

$$k_{eff} = \frac{\Delta \ln[IH]_o}{\Delta\tau}, \qquad (2)$$

Studying the polymer by method of differential scanning calorimetry we found its melting temperature 163° C.

Figure 1 shows the examples of curves of oxygen consumption during IPP oxidation on the presence of MBP. It is seen that the curves are characterized with very distinct induction periods. Figure 2 shows that the experimental data plotted in coordinates τ - $\lg[IH]_o$ are reliable straight lines, i.e. obey the formula (1). It is also seen that, and the data obtained from slopes of these lines are reliable, when the meanings of abscissas of the intersection points corresponding to critical antioxidant concentrations are less reliable.

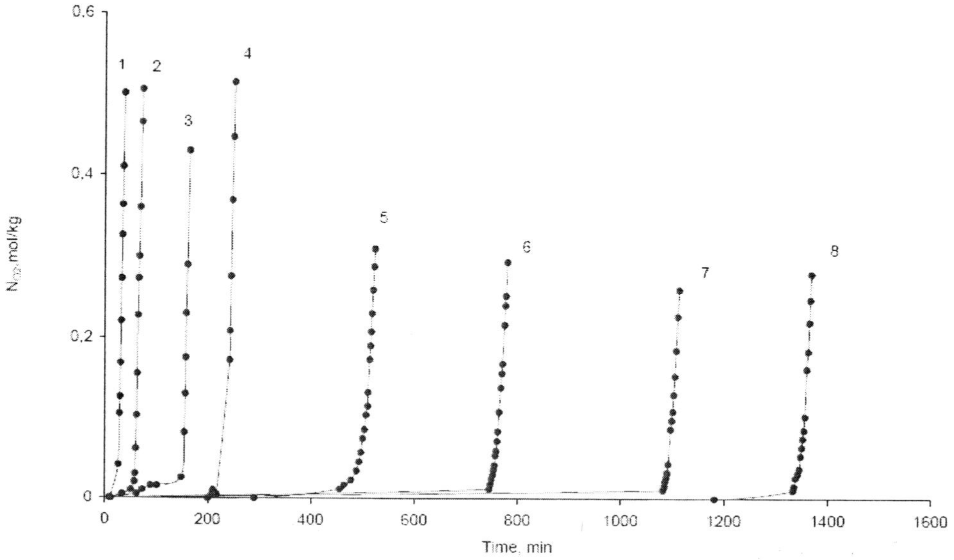

Figure 1. Oxygen consumption at oxidation of polypropylene: 210° C, MBP concentrations are 0,002 (1), 0,003 (2), 0,005 (3), 006 (4); 160°C, MBP concentrations are 0,0003(5), 0,0004 (6), 0,0006 (7), 0,00075(8) mol/kg , Oxygen pressure 300 mm Hg.

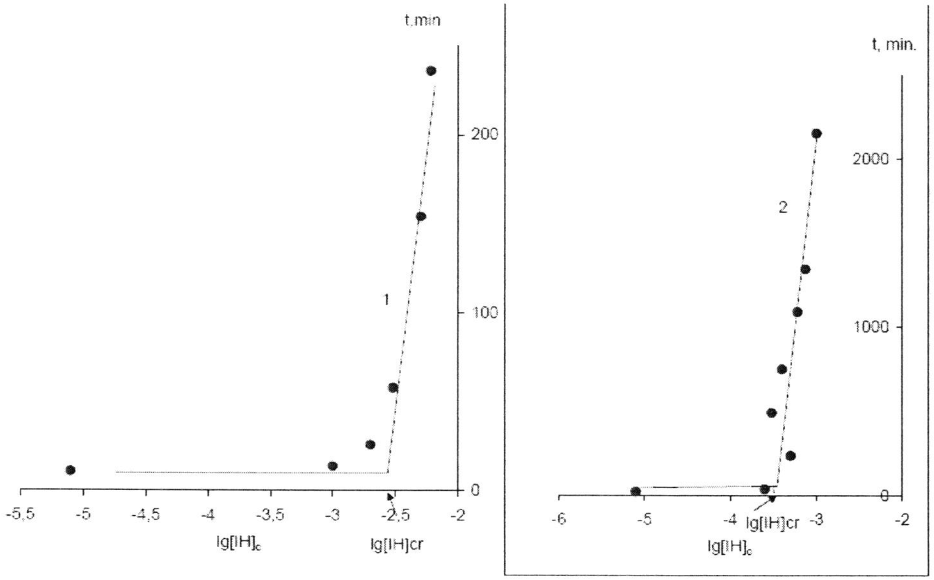

Figure 2. Dependencies of induction period of IPP oxidation on logarithm of initial antioxidant concentration. Temperatures: 210° C (1), 160°C (2).

Figure 3 shows the dependencies of critical concentration of MBP and the logarithm of the rate constant of its consumption on inverse temperature (1/T).

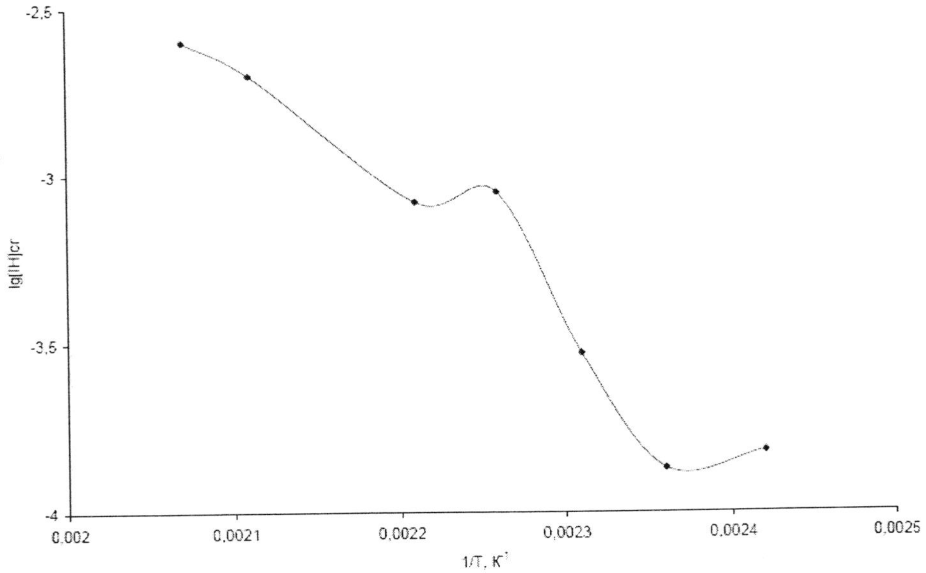

Figure 3. Dependencies of logarithm critical antioxidant concentration on oxidation temperatute in the coordinates of Arrhenius Law.

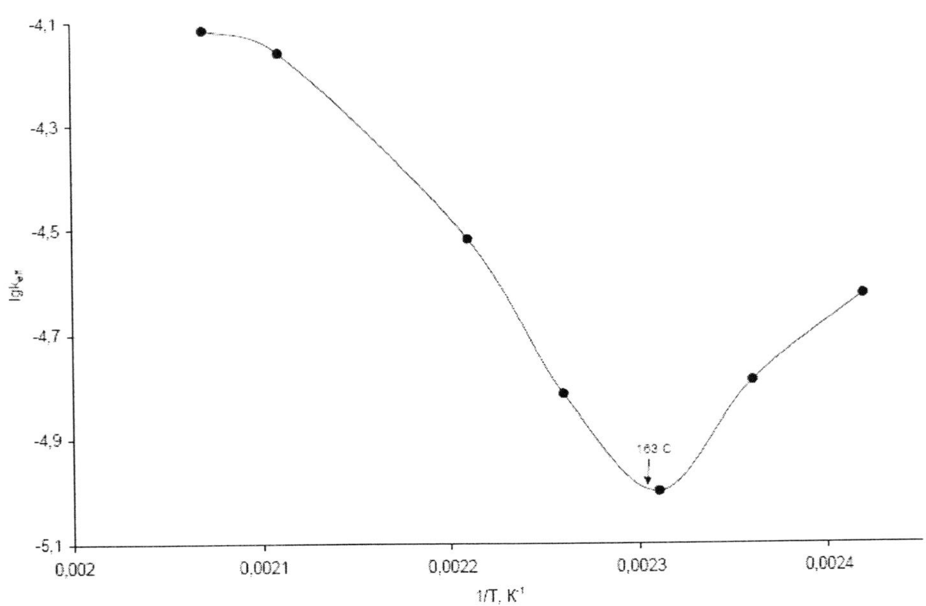

Figure 4. Dependencies of logarithm of apparent rate constant on oxidation temperatute in the coordinates of Arrhenius Law.

The dependence of critical concentration is presented by the curved line with maximum, when the rate constant in these (Arrhenius) coordinates consists of two wings meeting in the point corresponding to the melting temperature of IPP (163° C). Opposite to polyethylene both wings are curved lines, which does not make possible to calculate the activation energies of the process responsible for antioxidant consumption.

It must be specially noted that below the melting point of the polymer the rate constant increases with decreasing temperature. Top explain this phenomenon we must assume that the polypropylene melt contains complicated structures [5] the ratios of which change with temperature. Below the melting point the elements of disorder [6] present in the polymer are displaced by growing crystallites into non-crystallized elements of amorphous phase, the composition of which is varying with the temperature. The compounds dissolved in the polymer are present mainly in the sorption centers formed around these elements of disorder, and compression of these effects the rates of their interaction. As the result the rate constant increases 2.3 times at transition from 163° to 140°C. Another conclusion from our data is: the use of the simple Arrhenius formula for extrapolation over the melting temperature is incorrect. The authors are grateful to P.M. Nedorezova for synthesis of IPP on metallocen catalyst.

REFERENCES

[1] Nedorezova P.M., Aladyshev A.M., Savinov D.V., Klyamkina A.N., Optov V.A. Lementovskiy D.A., *Vysokomolecular. Soedin.* (Russian Polymer Science), 1991, A, V. 43, No 4, P. 595.

[2] Shlyapnikov Yu.A., Kiryushkin S.G., Mar'in A.P., *Antioxidative Stabilization of Polymers,* Taylor and Francis, London, 1996.

[3] Shlyapnikov Yu.A. in Developments in Polymer Stabilisation-5, G. Scott (Editor), *Applied Science*, London and New Jersey. 1978. P. 1.

[4] Shlyapnikov Yu.A., Kolesnikova N.N.. in: *Chemical Physics of Pyrolysis, Combustion and Oxidation,* Nova Science Publishers, Ed. A.A. Berlin et al., 2004. P. 111.

[5] Varga J., Menzel J., Sholti A., *Vysokomolek. Soedin.* (Russian Polymer Science) 1984, A, V. 26. P 2467.

[6] Shlyapnikov Yu.A., *Rusian Chemical reviews*, 1997, V. 66. No 11. P. 963.

In: Biochemical Physics Frontal Research
Editors: Sergei D. Varfolomeev et al., pp. 7-13

ISBN 1-60021-425-8

Chapter 2

KINETICS OF SULFUR VULCANIZATION OF 1,4- *CIS* -POLYISOPRENE BY DIFFERENTIAL SCANNING CALORIMETRY

N. N. Kolesnikova, O. A. Ostroukhova and A. A. Popov
N. M.Emanuel Institute of Biochemical Physics,
Russian Academy of Sciencis,
4, Kosygin str., Moscow 11991, Russia

ABSTRACT

The vulcanization of 1,4-cis-polyisoprene was studied by differential scanning calorimetry (DSC) under isothermal and non-isothermal conditions. On the basis of thermal characteristics obtained, the kinetic parameters of cross-linking (the induction period, maximum rate, and effective energy of activation) were determined.

Keywords: Differential scanning calorimetry, kinetics, vulcanization, polyisoprene.

As distinct from the conventional methods of studying the vulcanization of rubbers via determining the relationships between the kinetics and variable parameters that depend on a set of chemical reactions associated with cross-linking of macromolecules, DSC makes it possible to observe the process in progress by measuring its heat effects.

According to the data published, DSC is used in studies on the reactions of low-molecular-weight bi- and polyfunctional compounds that yield cross-linked polymer structures [1–3]. Data on using DSC in studies on the vulcanization of unsaturated rubbers are reported in few works [1–3].

The application of DSC for quantitative measurements relies on the fact that the heat released while a sample is heated in a calorimeter is an indication of the course of the reaction (on condition that the sample mass remains unchanged during the experiment) [6]. The experimental curves show the temperature or time dependences of the heat flow (dH/dt). The measurable parameters are the heat flow, i.e., the heat released per time unit (the DSC isotherm ordinate) and the amount of heat released (ΔH).

The DSC principal assumption is that the chemical reaction rate at any given moment is directly proportional to the rate of heat release:

$$d\alpha/dt = dH/dt \cdot 1/\Delta H_0,$$

where $d\alpha/dt$ is the reaction rate, dH/dt is the heat released per time unit or the heat flow, and ΔH_0 is the reaction heat determined from the peak area between the DSC curve and the baseline.

The degree of conversion (cross-linking) at the moment **t** is determined as a ratio between the heat released by this moment and the total heat of the reaction:

$$\alpha = \Delta H_t / \Delta H_0.$$

The aim of this work is to investigate the feasibility of using DSC to study the sulfur-accelerated vulcanization of rubbers using 1,4-*cis*-polyisoprene as an example.

In the work, we used synthetic isoprene rubber (IR-3) (the content of 1,.4-*cis*-links 92–99%, $M_\eta \approx 1 \cdot 10^6$). The vulcanizing group used per 100 wt u. rubber consisted of 2 wt u. sulfur, 1 wt u. sulfonamide C, 5 wt u. zinc oxide, and 1 wt u. stearic acid. The rubber was mixed with the vulcanizing group ingredients in a laboratory mill at room temperature. The vulcanization was studied with the use of a DSM-2m differential scanning microcalorimeter. The temperature and heat flow calibration was performed against standard samples of indium (T_{melt} = 156.6°C; ΔH_{spec} = 28.4419 J/g). The mass of the sample under study was ~ 25 mg. Under isothermal conditions, the sample was heated to the preset temperature at a heating rate of 64 deg/min. Then, isotherms were recorded. The time during which the thermal stability was attained was 90 sec. The DSC isotherms were recorded at temperatures from 140 to 170°C. Under non-isothermal conditions, the samples were heated at a rate of 2, 4, 8, or 16 deg/min. The temperature dependences were obtained at 40 to 250°C. The baseline was preset before each scanning experiment using two empty containers at a similar heating rate. The reaction was considered completed when the DSC curve reached the baseline. Weight losses of the samples after the experiments were insignificant – no more than 0.5% of the initial sample mass. The amount of heat released was determined from the peak area between the DSC curve and baseline. The accuracy of measurements was ±1 °C for the temperature and ±5% for the heat of cross-linking.

The degree of cross-linking of samples heated in the calorimeter was determined from their equilibrium swelling in toluene.

A preliminary study showed that heating of the mixture of IR and the vulcanizing group at a preset rate or constant temperature yields an exothermal peak of cross-linking; whereas heating of IR in a calorimeter in the absence of the vulcanizing group yields a monotonic curve. The thermograms of samples obtained at various heating rates are shown in Figure 1. The vulcanization stage is initiated at the temperature (T_{init}) of completion of the induction period; this is in agreement with the concept about the mechanism of vulcanization [7]. It is evident from Fig. 1 that an increase in the scanning rate results in shifting the peak of cross-linking to the region of higher temperatures.

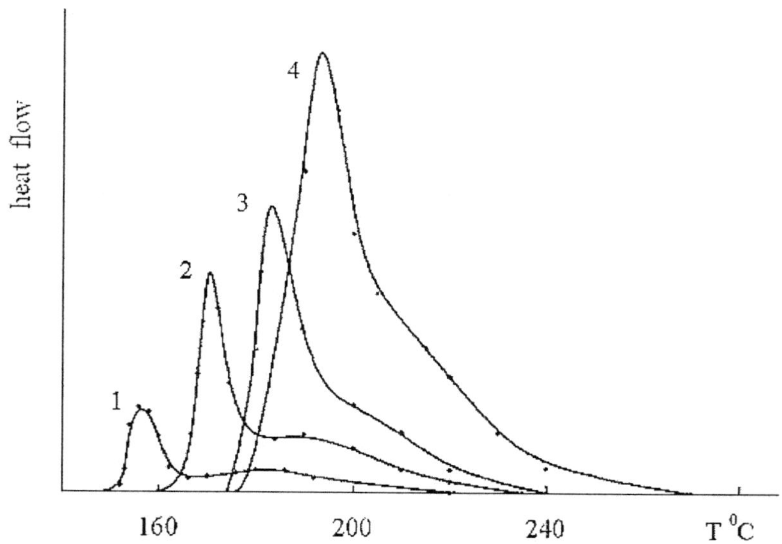

Figure 1. DSC thermograms of cross-linking IR for various scan rates: 2 (1), 4 (2), 8 (3) и 16 (4) deg/min.

From the thermograms obtained, the temperature of initiation (T_{init}) and the maximum rate (T_{max}) of cross-linking as well as the heat effect of the reaction at different scanning rates were determined. The total heat was calculated by the formula including the rate of heating of the sample,

$$\Delta H_0 = 1/\beta \int_{T_1}^{T_2} dH/dt \cdot dT,$$

where β is the heating rate.

The temperature and specific heat of IR cross-linking at various scanning rates are listed in Table 1. It is evident that the reaction heat increases slightly as the scanning rate is increased. At a high heating rate (16 deg/min), the specific heat of cross-linking decreases; the effect may be caused by the thermal decomposition of the bonds formed or by other secondary reactions that contribute to the total heat effect of the process.

Table 1. Characteristics of IR cross-linking at various scanning rates

β, deg/min	$T_{init.}$, °C	$T_{max.}$, °C	$\Delta H_{spec.}$, J/g
2	149	156	35
4	160	170	39
8	173	183	42
16	177	192	36

The shift in the temperature of the peak cross-linking of rubber to the region of higher temperatures with increasing the heating rate may be attributed to the influence of the molecular mobility on the chemical process. At a low heating rate, there is enough time for the chains to reach the states favorable for cross-linking. At a high heating rate, these states are attained later, at higher temperatures.

From the temperature dependences of IR cross-linking proceeding under isothermal conditions at various temperatures (Figure 2), it is evident that the higher the temperature, the sooner the cross-linking is initiated and the maximum rate is achieved.

From the peak areas of the isotherms, we determined the specific heats of cross-linking; their values are listed in Table 2. According to data shown in Table 2, the heat of cross-linking proceeding under isothermal conditions varies slightly with increasing the temperature; however it remains lower than the maximum heat of non-isothermal cross-linking. This is evidence for an incomplete cross-linking of rubber under isothermal conditions.

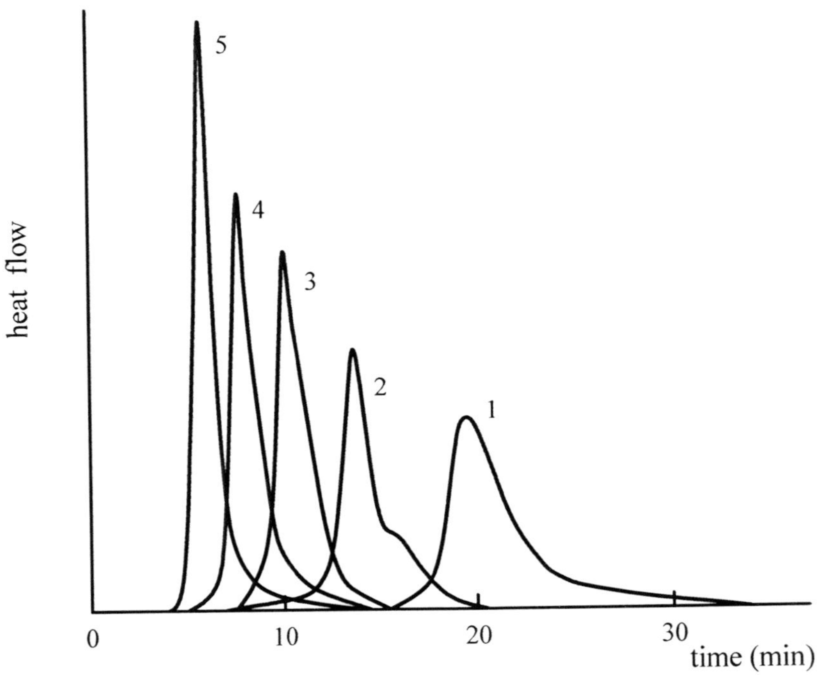

Figure 2. Isothermal DSC curves of cross-linking IR at 140 (1), 145 (2), 150(3), 155(4) and 160 (5) ^{0}C.

Table 2. Characteristics of isothermal IR cross-linking

T, ^{0}C	τ, min.	$\Delta H_{spec.}$, J/g	$V_{max.} \cdot 10^3$, s^{-1}
140	15,5	14,0	1,0
145	7,5	12,0	1,0
150	7,0	13,0	2,0
155	5,0	13,0	3,0
160	4,0	13,0	3,5

The degree of cross-linking achieved by the time **t** was determined as $\alpha = \Delta H_t/\Delta H_0$, where ΔH_0 is the heat of cross-linking of rubber under non-isothermal conditions at a heating rate of 4°/min. Figure 3 shows the kinetic curves of the IR vulcanization proceeding under isothermal conditions at various temperatures. In all the cases, we observe a drastic increase in the cross-linking rate after the end of the induction period (τ). The induction period (the absence of a recordable heat effect) involves melting and interactions of the vulcanizing group components; these processes are associated with an insignificant total heat effect. The quasi-stationary segment of the kinetic curve corresponds to the period of formation of cross-links between macromolecules. From the kinetic curves, we determined the maximum rate and the effective energy of activation of IR cross-linking with the vulcanizing group chosen. The values of the reaction rates (V_{max}) are listed in Table 2.

On the basis of the literature data [2] and measurement results on the heat release, the kinetic model of cross-linking of rubber was constructed. The period of cross-linking may be described by the first-order equation:

$$d\alpha/dt = k\,(1 - \alpha),$$

where k is the reaction rate constant. With regard to the temperature dependence of k, we get the equation for the rate of the reaction proceeding under isothermal conditions

$$d\alpha/dt = A \cdot \exp[-E_a/RT](1 - \alpha),$$

where A and E_a are the pre-exponential multiplier and energy of vulcanization, respectively; R is the gas constant equal to 8.314 J/mol·K; and T is the temperature, K.

The effective energy of activation of cross-linking was calculated in different ways: by the Arrhenius equation

$$\ln (d\alpha/dt) = \ln (dH/dt \cdot 1/\Delta H_0) = \ln A - E_a/RT + \ln (1 - \alpha)$$

and from the shift in the temperature of the cross-linking exotherm peak depending on the heating rate β; it is calculated by the Ozawa equation [8].

$$\ln\beta = \text{const} - 1.052\, E_a/RT_{max}$$

The value of $E_{a\ eff}$ determined from the maximum rates of IR cross-linking at various temperatures under isothermal conditions (91 kJ/mol) is similar to that calculated from the shift in the temperature of the cross-linking peak at various heating rates under non-isothermal conditions (88 kJ/mol). Note that the calculated energy of activation and the temperature coefficient of the reaction, which is equal to 2, are consistent with the literature data on the sulfur-accelerated vulcanization [7].

The fact that heat released during heating the samples is determined by cross-linking is verified by the data on equilibrium swelling of samples heated in the calorimetric cup during different times at the same temperature.

Figure 4 shows the dependences of the IR cross-linking degree (kinetic curves) (155°C) determined by the methods of DSC and equilibrium swelling (in the latter case, the value 1/Q, which is proportional to the cross-linking degree, is determined). As is seen from Figure 4, the shapes of the cross-linking curves are similar.

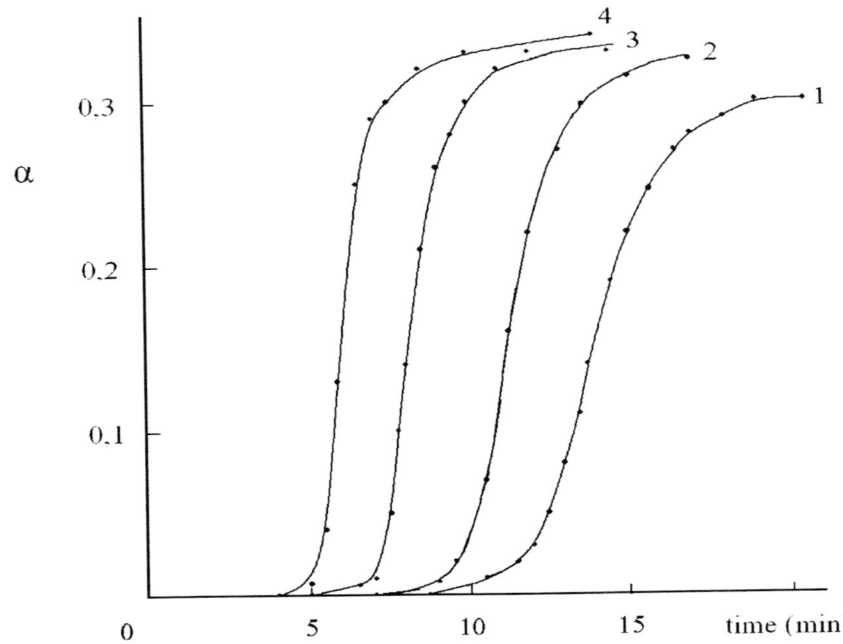

Figure 3. Kinetic curves of cross-linking IR at 145 (1), 150 (2), 155 (3) and 160 (4) ^{0}C.

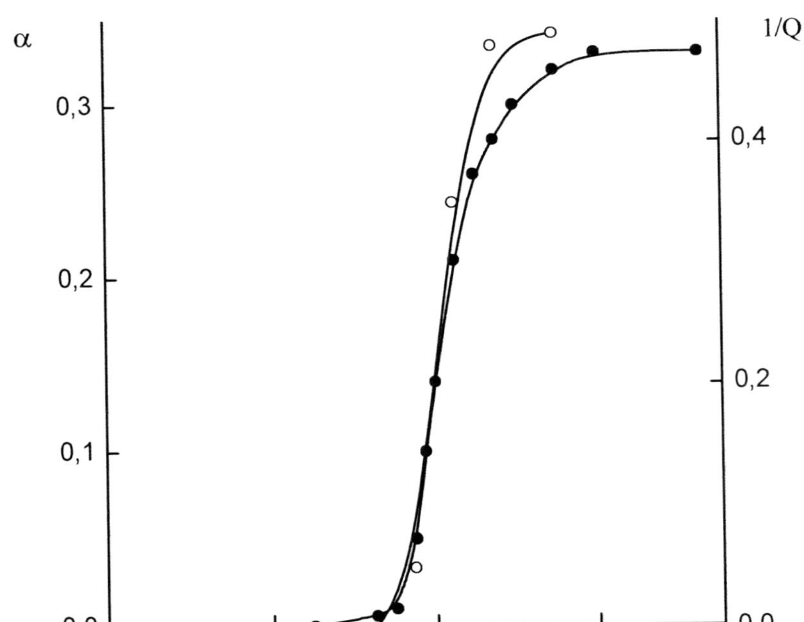

Figure 4. Dependence of cross-linking degree IR on time determined by the methods of DSC (●) and equilibrium swelling (○) at 155°C.

Thus, differential scanning calorimetry is a rapid and convenient method of studying the whole process of sulfur-accelerated vulcanization including the induction period. The kinetic parameters determined in this work (duration of the induction period, rate of vulcanization, and effective energy of activation) are consistent with the literature data.

REFERENCES

[1] H. Ishida, Y. Rodriguez, *Polymer,* Vol.36, N 16, pp. 3151-3158, 1995.

[2] F.Y.C. Boey, F.Y.C. Qiang W., *Polymer*, Vol. 41, N 6, pp.2081-2094, 2000.

[3] M.J. Starink, *Thermochimica Acta*, Vol. 404, N 1, pp.163-176, 2003.

[4] S.-H. Chough, D.-H. Chang, *Journal of Applied Polymer Science*, Vol. 61, pp. 449-454, 1996.

[5] P. Simon, A. Kusma, *Journal of Thermal Analysis and Calorimetry*, Vol.56, N 3, pp. 1107-1113, 1999.

[6] V.A. Bershteyn, V.M. Egorov, *"Differential scanning calorimetry for physicochemistry of polymers"*, Leningrad, Chimiya, 1990, p.221.

[7] B.A. Dogadkin, A.A. Donczov, V.A. Shershnev, "Chemistry of elastomers", Moskva, *Chimiya,* 1981, 373pp.

[8] T. Ozawa, *Journal of Thermal Analysis,* Vol.9, pp.369-370, 1976.

In: Biochemical Physics Frontal Research
Editors: Sergei D. Varfolomeev et al., pp. 15-19

ISBN 1-60021-425-8

Chapter 3

INFLUENCE OF ELECTRIC AND MECHANICAL FORCES ON THE OXIDATION KINETICS AND ELECTROCONDUCTIVE POLYISOPRENE COMPOSITION PROPERTIES

V. E. Gul', I. A. Kirsh, E. V. Kvasnikova, D. U. Zabulonov and M. V. Soina

Moscow State University of Applied Biotechnology,
33, Talalikhin str., Moscow

ABSTRACT

The investigation results of the simultaneous effect of electric current and tensile deformation on the oxidation processes and electroconductive polyisoprene composition (EPIC) properties are presented in the study.

The effects of the EPIC oxidation process under the conditions of the complex factor action consisting of the gaseous product emission and decrease of the oxidation rate at the final stage of the experiment were disclosed.

It was found out that the change in the electric resistance with time of EPIC aging was characterized by three sections: decrease, constant value area and increase of the electric resistance. It was demonstrated that the values of the electric resistance and duration of the above sections depended on the deformation temperature and extent.

INTRODUCTION

When using articles of polymer materials the reactions of the air oxygen interaction are of a great significance for holding their operational characteristics. These reactions are to a great measure affected by the service temperature and action of various forces, chiefly, mechanical and electric. In particular, articles of electroconductive polymer compositions are subjected to such actions.

At present the demand for electroconductive polymer materials applied as various kinds of heating elements, antistatic coatings, electrode circuits, grounding devices, etc. is increasing. The continuously increasing range of the articles attaches a particular importance to the problem of their serviceability maintaining.

One of the primary factors limiting the articles' serviceability is the process of the chemical interaction of a polymer composition with air oxygen. In addition, the articles are exposed to the action of such factors as temperature, electric current and mechanical loads that affects to a certain measure the oxidation processes of the polymer matrix.

The previous studies [1, 2] on investigating the electric current influence on the oxidation of electroconductive polymer compositions demonstrated the increase in the rate of the oxidation process upon electric current flowing. However, these studies took no account of mechanical effects. Therefore, this study is devoted to the investigation of the simultaneous influence of electric current and tensile deformation on the oxidation process and EPIC properties.

EXPERIMENT

The object of the investigation was synthetic polyisoprene rubber-based composition SKI-3 (GOST 14925-79) containing commercial carbon of the П-367-Э grade in the amount of 50 mass parts. The polyisoprene composition comprised the following components: sulfur, sulfenamide Ц, zinc oxide, stearic acid.

The process of the specimen manufacture was performed in several stages: mixture preparation, component blending and curing. Blending was performed in the laboratory mixer of the "Banbury" type. Upon emerging from the mixer the mixture was squeezed between rolls. Curing was carried out in the steam-heated hydraulic press at a curing temperature of 143°C.

The plant for accelerated polymer aging was used as the method to investigate the oxidation process under the effect of temperature, electric current and deformation. The plant's principle of operation consists of registering a pressure change in the system where the specimen is tested. Since this plant [2] was previously used to study the thermoelectrooxidation destruction of electroconductive polymer compositions without any mechanical effects then some changes were made enabling the oxidation process to proceed under the simultaneous effect of current and deformation. In compliance with the task to be solved the design of the reaction dish and clamp-furnished base for specimen securing was changed.

The electric resistance was measured by means of a digital versatile instrument C0234 with the specimen being not removed from the reaction dish.

The following methods were also applied in the study: gas chromatography, electronic microscopy.

Results and Discussions

In the process of EPIC oxidation the pressure change in the system was registered during the experiment period. Since the oxidation was to be performed under the multiple-factor action, then the investigations were conducted under the conditions both of the separate factors' action and also of their combination. That allowed to note some features of these factor effect on the oxidation processes.

The investigations of the EPIC oxidation process at a various extent of the tensile deformation demonstrated that under the conditions of the electric current flow through the composition specimens (Figure 1) the original area of the kinetic oxidation curves was characterized by an area of the gaseous product emission (not absorbed by absorbents). This area is unavailable for the oxidation process performed under the same conditions, but without any current (Figure 2). Under these conditions an induction period is observed in the curves' initial area. The performed analysis of the gaseous substances emitted at the initial stage (as a result of oxidation under the electric current flow) showed the presence of methane, hydrogen and hydrocarbon mixture.

It was found out as a result of this investigation that with the increase of the deformation extent the gaseous product emission time decreased and the maximum value of the oxidation rate and pressure change in the system increased. That is accounted for by the action of mechanical load, in particular, tensile deformation that contributes to the intensification of the oxidation process. The similar regularity was observed in EPIC oxidation without any current, however, in this case the induction period decreased.

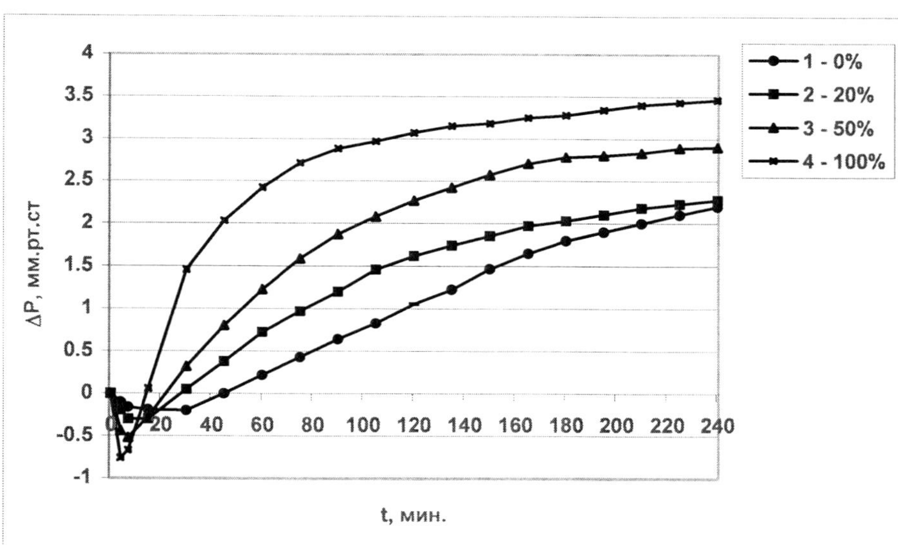

Figure 1. Change of pressure in system from time of ageing at T=373K, *U=60V* and a different degree of deformation.

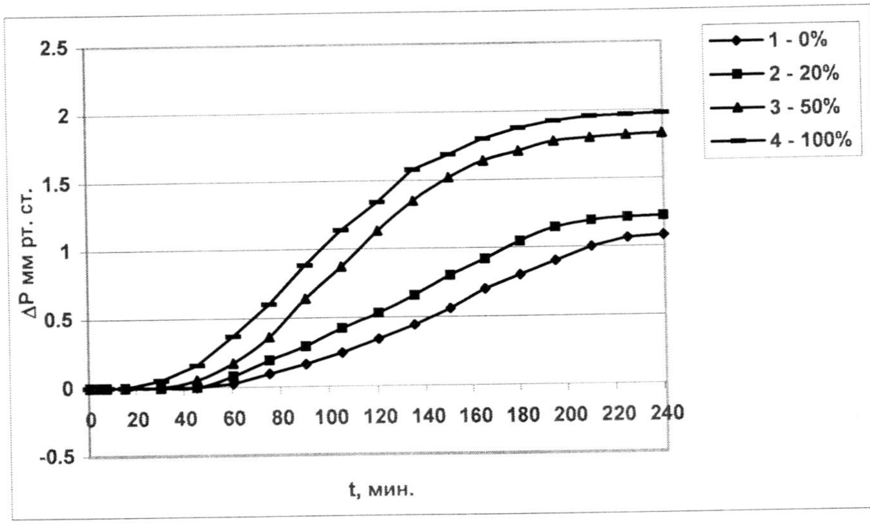

Figure 2. Change of pressure in system from time of ageing at T=373K, $U=0V$ and a different degree of deformation.

It is reasonable to point out that the maximum rate of the oxidation process without current increases with deformation to a lesser extent than for the aging process under current that is caused by its direct effect. The maximum value of the pressure change in the system is higher for current-aged specimens.

In the final area of the curves upon the EPIC specimen deformation stable values of the pressure change in the system are observed. The observed effect of the oxidation rate decrease is not related to the electric current influence since it was revealed without it as well if the specimen was subjected to deformation in this case. It was found out that the time of this effect manifestation depended on the deformation extent and experiment temperature. For instance, the time of the gaseous product emission decreases with the temperature, the oxidation rate increases and the pressure stabilization effect is observed later. At sufficiently high temperatures (140°C) this effect within the testing time interval was absent that is related to the formation of additional active centers that assist in intensifying the oxidation processes.

For the purpose of forecasting the time of the composition serviceability holding the electric characteristics were investigated under the conditions of the complex factor effect. Figure 3 shows the results of the electric resistance change depending on the EPIC aging time.

At the commencement of the experiment a considerable decrease in the electric resistance is observed. That is related to the electric current flow since when conducting the experiment on EPIC aging under the same conditions, but without current a negligible decrease in R values was noted. The electric current flow during the oxidation process causes the current-conducting path reformation that is a result of an abrupt decrease in the electric resistance. This assumption is confirmed by the experiments conducted using the method of the electronic microscopy.

The decrease in the electric resistance of the electroconductive polymer composition is also affected by the tensile deformation. For instance, the value of the electric resistance decreases with the deformation that is related to the orientation of the soot structure elements along the tensile axis. The values of the electric resistance behind the area of the electric resistance decrease in curves lg R = f(t) do not in practice change in the case of 20% and 50% deformation and upon the 100% deformation the increased values of the electric resistance in the final area of the experiment is noted.

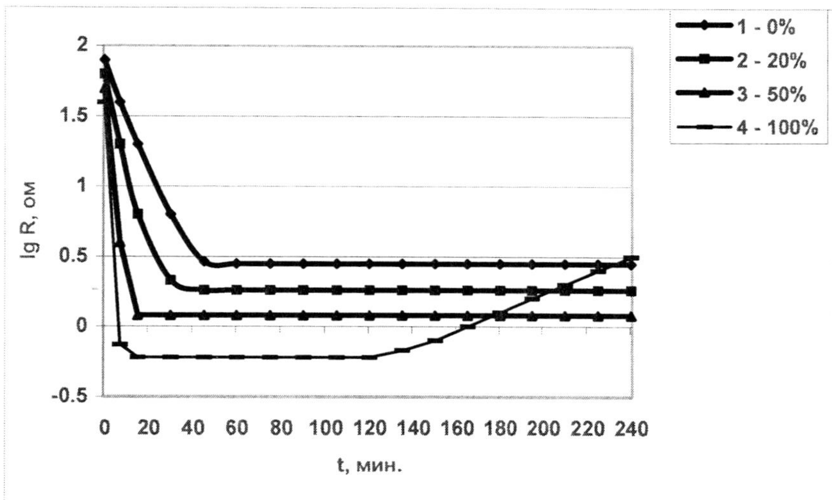

Figure 3. Change of electric resistance from time of ageing at T=373K, *U=60V* and a different degree of deformation.

The area of the electric resistance increase with the temperature was also observed at other deformation degrees and it occurred at a shorter aging time. For instance, at a temperature of 140°C it was observed under all the deformations extents and even without it. That is explained by the fact that the destruction processes that are accompanied with the current-conducting path destruction intensify with the temperature.

The electric resistance increase can further result in loss of the electroconductive polyisoprene-based article serviceability. Therefore, the theoretical service life was computed on the bases of the obtained data using the mathematical description programs. The results of the performed serviceability computation found practical application in choosing a composition for the anode grounding system: rust-protection element for gas pipelines.

A design of the re-usable device for heating packed foodstuff was developed by the results of the oxidation investigations under conditions of the complex factor action (electric current, deformation and heating) and the period of its safe operation was established.

REFERENCES

[1] Gul' V.E., Ryjenkova I.P. New Method for Thermoelectrooxidation Investigations of Electroconductiv Polymer Compositions. *Polymer Testing.* 13, 1994, p. 181-187.

[2] Gul' V.E., Mokeeva E.V. Influence of Electric Current on the Oxidation Process of Thermostable Polymer Composition. *Plastic Masses,* No. 1, 1998, p. 5-8.

In: Biochemical Physics Frontal Research
Editors: Sergei D. Varfolomeev et al., pp. 21-33
ISBN 1-60021-425-8

Chapter 4

ANALYSIS OF A REGULATORY PEPTIDE FROM THE BOVINE EYE LENS: PHYSICOCHEMICAL PROPERTIES AND EFFECT ON CATARACT DEVELOPMENT *IN VITRO* AND *IN VIVO*

***M. S. Krasnov, E. P. Gurmizov*[*]*, V. P. Yamskova and I. A. Yamskov*[**]**

N. K. Kol'tsov Institute of Developmental Biology,
Russian Academy of Sciences, 26 Vavilova str., Moscow 119991, Russia
[*]Helmholts Research Institute of Eye Diseases,
RF Ministry of Public Health and Social Development
14/19 Sadovo-Chernogryazskaya str., Moscow 105062, Russia
[**]A.N. Nesmeyanov Institute of Organoelement Compounds,
Russian Academy of Sciences, 28 Vavilova str., Moscow 119991, Russia

ABSTRACT

A regulatory peptide (RP) biologically active at ultralow doses (10^{-12} mg/ml) was isolated from the bovine eye lens and studied with respect to amino acid composition and secondary structure. This RP was shown to have a protective effect on eye cells in experiments on chemically induced cataract in amphibian and mammalian lenses cultured *in vitro* and traumatic cataracts development *in vivo* in rats. Using an immunohistochemical method, the RP was localized in the intercellular space of the lens epithelium.

Keywords: Regulatory peptides, eye lens, ultralow doses, cataract

INTRODUCTION

The purpose of this study was to analyze physicochemical properties of a new regulatory peptide (RP) isolated from the bovine eye lens and to assess its effect on cataract development in vitro and in vivo. The RP belongs to the group of regulatory molecules that exhibit biological activity at ultralow concentrations, influencing the course and direction of major biological processes such as cell adhesion, migration, proliferation, and differentiation [Yamskova and Reznikova, 1991; Yamskov et al., 1999; Krasnov et al., 2000, 2003a, 2003b]. In particular, these RPs improve the viability of cells and provide for the maintenance of their morphogenetic reactions under conditions of in vitro culture [Krasnov et al., 2003a]. Our previous data show that the biological activity of these RPs is not species-specific [Krasnov et al., 2003a, 2003b; Margasyuk et al., 2005b].

Cataract is a widespread human eye disease. The loss of transparency of the eye lens may be caused by different factors but is always associated with disturbances in the arrangement of lens fibers and conformation of lens proteins. Today, cataracts are mainly treated surgically, by replacing the opaque lens with a plastic artificial lens [Kopaeva, 2002]. Nevertheless, the development of pharmacological agents maintaining and improving eye lens transparency is an important problem in modern ophthalmology. In this study, we tested the RP from the bovine eye lens as a prospective anticataract agent on the experimental models of in vitro cultures of vertebrate eye lenses and cataracts development in vivo after trauma in rats.

MATERIAL AND METHODS

Wistar rats (males and females weighing 210–230 g) and mature *Xenopus laevis* and *Rana temporaria* frogs were from the vivarium (aquarium) of the Kol'tsov Institute of Developmental Biology. Bovine eye lenses (from freshly enucleated eyes) were from the slaughterhouse of the Taganka Meat-Packing Plant. All reagents were of chemical purity grade. Solutions were made in deionized water (15 MΩ) from a MilliQ column.

To prepare the initial extract, whole bovine lenses were incubated in a saline (0.15M NaCl; 1 mM $CaCl_2$; 1mM HEPES) at 4–8^{0}C for 2 h. A total of 150 lenses were used for this purpose. The extract was then centrifuged at 3000 *g* for 10 min to remove cells and separated according to the procedure tested in our previous studies, which involved two rounds of protein precipitation with ammonium sulfate and isoelectrofocusing (IEF) of the supernatant fluid in a sucrose density gradient with ampholytes [Yamskova and Reznikova, 1991]. Fractions were detected spectrophotometrically at 280 nm, and pH range was determined for each fraction. We collected the acidic fraction (pH < 3.0), which was dialyzed, concentrated, and further separated by nondenaturing electrophoresis in 7.5% PAAG [Hames, 1990]. The zone corresponding to a peptide with $R_f \approx 0.90 \pm 0.05$ was cut out of the gel and incubated in several small volumes of water to elute the peptide. The eluates were pooled, dialyzed several times against water, and lyophilized. The resultant preparation was analyzed by several methods.

Protein concentration in solutions was determined by the colorimetric method with copper bicinchoninate [Smith et al., 1985].

The dependence of biological activity on the dose of RP preparation was estimated in the cell adhesion test developed for biological indication of proteins belonging to this group [Yamskova and Reznikova, 1984].

SDS-electrophoresis in 15% PAAG [Laemmli, 1970] was performed in a Mini-Protean 3 cell (BioRad, Italy). To estimate parameters of the secondary structure of test proteins, their circular dichroism spectra were recorded in a Jasco 720 dichrograph (Japan) equipped with a computer data processing system. The spectra were analyzed using the k2d program [Venyaminov, Yang, 1996]. The amino acid composition of RP was determined in a Hitachi-835 amino acid analyzer (Japan).

Antisera for immunohistochemical analysis were obtained by immunizing Chnchilla rabbits (≈ 1 kg) with the RP preparation, using a conventional procedure [Gosling, 2000]. Serum samples were tested for the presence of specific antibodies by means of ELISA [Engvall and Perlman, 1971]. An indirect immunofluorescent method [Engelhardt, 1968] with these primary antibodies and FITC-labeled secondary antibodies to rabbit IgG (Sigma, United States) were used for localizing the RP in cryostat sections (10 μm) of the rat eye lens fixed with chilled ethyl alcohol. The primary and secondary antibodies were diluted 1:10 and 1:35, respectively. Control sections were treated with (1) secondary antibodies alone or (2) a nonimmune rabbit serum.

The amphibians were anesthetized with 0.1% MS-222 solution and decapitated. The rats were anesthetized with diethyl ether and decapitated. Amphibian lenses were cultured in plastic 24-well plates, 1 ml of the medium per well, at 22°C; rat and bovine lenses were cultured in glass vessels containing 5 and 100 ml of the medium, respectively, at 37°C for 3–8 days. The vessels were sealed with Parafilm and aluminum foil.

The development of cataract *in vitro* was induced by supplementing the culture medium with hydrogen peroxide or calcium chloride. In experiments with rat, bovine, *R. temporaria,* and *X. laevis* lenses, the respective concentrations of these agents were 0.5, 40, 2, and 0.5 mM (hydrogen peroxide) and 15, 20, 10, and 2.5–3 mM. These concentrations were chosen taking into account previous data on the optimal correlation between the time of treatment and the degree of lens opacification [Cui and Lou, 1993; Lou et al., 1995; Krasnov et al., 2005]. The RP from the bovine lens was added simultaneously with the cataract-inducing agents to a final concentration of 10^{-12} mg/ml. The degree of lens opacification was estimated visually and by measuring optical density at 340, 405, 490, 550, and 630 nm with an MR700 microplate reader (Dynatech, United States).

The rats were anesthetized with diethyl ether and 1% tropicamide solution was instilled once in each eye. In experiments on the development of traumatic cataract in Wistar rats, the anterior segment of the lens was punctured to a specified depth with a needle (0.45 mm in diameter) inserted through the cornea. Two series of experiments were performed, each with two groups of six rats (the control and test groups). In the first series, the lens was punctured to a depth of 5.1 mm from the cornea surface, so that the needle passed through the lens nucleus to the posterior cortical layers. In the second series, the puncture was 3.4 mm deep, so that only the anterior cortical layers of the lens were damaged. In both series, eye injuries in rats of the test group were treated with the RP preparation (10^{-12} mg/ml) instilled once a day at a dose of 50 μl, and rats of the control group received the same dose of saline. The experiment continued for seven days.

Experimental data were processed by methods of variation statistics. Student's *t*-test was used to estimate significance of differences.

RESULTS AND DISCUSSION

The bovine lens RP was isolated and purified using the procedure developed for proteins of this group [Yamskova et al., 1977; Yamskov et al., 1999]. At the last stage, the procedure involved nondenaturing electrophoresis in 7.5% PAAG [Hames, 1990]. According to the results of SDS-electrophoresis in 15% PAAG, the apparent molecular weight of the resultant peptide was less than 3.5 kDa (Figure 1). As the fraction containing this peptide has a strong negative charge (pI < 3.0), its true molecular weight may be slightly greater. The amino acid composition of the RP (Table 1) is characterized by the prevalence of dicarboxylic amino acids, glycine and serine; there is also a considerable number of hydrophobic amino acid residues, whereas the proportion of aromatic amino acid and methionine residues in insignificant. The secondary structure of the RP molecule was inferred from circular dichroism spectra. As follows from Table 2, this molecule contains extensive β-sheet regions (about 65%), most of them being antiparallel, whereas the content of α-helices is low (about 7%). The remainder of the polypeptide chain has no definite folding pattern (a stochastic globule).

Table 1. Amino acid composition of RP isolated from the bovine eye lens

Amino acids	% (weight)
Ala	7,4±0,5
Asp/Asn	10,7±1,0
Arg	4,8±0,5
Val	5,4±0,5
His	6,1±0,5
Gly	6,6±0,5
Glu/Gln	21,4±1,5
Ile	3,1±0,5
Leu	4,7±0,5
Lys	4,6±0,5
Met	1,8±0,5
Pro	5,0±1,0
Ser	7,0±1,5
Tyr	2,9±0,5
Thr	5,7±0,5
Trp	0
Phe	2,8±0,5
Cys	0
Total	100±1,0

Table 2. Parameters of secondary structure of RP isolated from the bovine eye lens

Elements of secondary structure	Content in RP molecules, %
α-Helices	6,8±0,5 %
Antiparallel β-sheets	43,4±1,0 %
Parallel β-sheets	5,2 ±0,5%
β-turns	18,5±1,0 %
Random coil	32,8±1,0 %

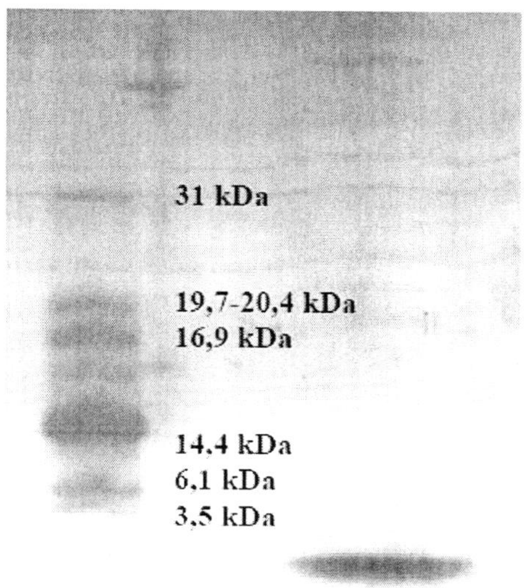

Figure 1. SDS-electrophoresis of RP from the bovine eye lens in 15% PAAG.

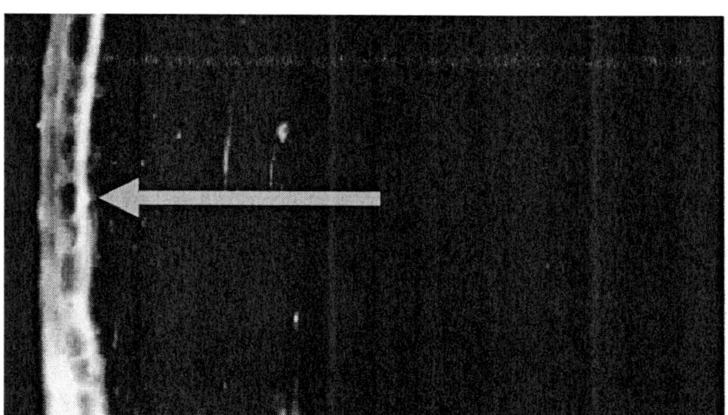

Figure 2. Immunochemical localization of RP from the bovine eye lens in the rat lens epithelium. The arrow indicates RP accumulation near the basal membrane. Magnification: ocular 10×, objective lens 20×.

In cell adhesion tests, the highly purified RP fraction (obtained after electrophoresis) was biologically active at a concentration of 10^{-12} mg protein/ml. It was used as an antigen for raising specific rabbit antibodies to this RP. By means of immunohistochemical analysis with these antibodies, the RP was localized in the intercellular space of the rat lens epithelium (Figure 2), mainly at the basal membrane.

The effect of lens RP on the development of cataract was tested on frog (*Rana temporaria* and *Xenopus laevis*), rat, and bovine eye lenses cultured *in vitro* for eight days. Lens opacification was induced by changing calcium ion concentration in the culture medium or by adding hydrogen peroxide. These agents were chosen for several reasons. The mechanism of lens damage by hydrogen peroxide, which leads to opacification of the cortical lens layers, involves activation of lipid peroxidation in membranes of lens cells, which takes place in the cases of complicated, traumatic, and certain senile cataracts in humans [Bhuyan et al., 1984; Bhatnagar et al., 1993]. When a cataract develops under the effect of oxidants, the diameter and weight of the lens increase, with the increase in diameter preceding opacification [Bhatnagar et al., 1993]. According to these authors, lens transparency in the presence of oxidants exponentially decreases with time, and this process develops primarily in the lens cortex; the lens nucleus is affected at later stages.

The cataract-inducing effect of calcium ions is explained by changes in the functioning of Ca^{2+}-ATPase systems in lens cells [Bhatnagar et al., 1995; Siew and Bettelheim, 1996]. The excessive input of calcium ions disturbs the arrangement and shape of lens fibers, and, as a result, structures that scatter light appear in the lens during cataract development. The lens nucleus is often opacified under such conditions. Hence, this model may be regarded as an analog of the nuclear form of senile cataract in humans [David and Shearer, 1984; Tang et al., 2003]. Thus exposure to these two chemical agents leads to the development of cataract in the forms that are most often recorded in ophthalmological practice.

We performed five series of experiments with eye lenses incubated in the following media (8–10 lenses from each animal species per variant): (1) medium 199, (2) medium 199 + H_2O_2, (3) medium 199 + H_2O_2 + RP solution (10^{-12} mg/ml), (4) medium 199 + $CaCl_2$, and (5) medium 199 + $CaCl_2$ + RP solution (10^{-12} mg/ml).

The results of these experiments are shown in Figures 3–7.

In series 1, eye lenses cultured in the medium without serum and cataract-inducing agents remained transparent and retained normal morphology throughout the observation period (Figures 3–7). In series 2 and 4 (exposure to cataract-inducing agents alone), significant lens opacification was observed on days 3–4. In series 3 and 5 (exposure to cataract-inducing agents combined with RP), opacification was insignificant, the lenses became slightly "frosted."

In amphibian lenses, exposure to hydrogen peroxide resulted in opacification of cortical layers, whereas the nucleus remained transparent; in the variant with RP solution, the cortical layers were opacified only slightly (Figure 3). Under the effect of calcium ions, only the nucleus was opacified, whereas the cortical layers remained transparent or, in a few cases, translucent; when RP was added, the lens nucleus also remained transparent (Figures 4, 5).

In bovine lenses, exposure to either factor resulted in medium-degree cortical opacification on days 3–4. When the medium was supplemented with the RP solution, the degree of opacification was markedly lower (Figure 6). Rat lenses exposed to hydrogen peroxide manifested opacification of both the nucleus and, to a greater extent, cortical layers

(Figure 7a). This agrees with the results obtained by Cui and Lou (1993) with rat eye lenses cultured in the medium containing 0.5 mM hydrogen peroxide, in which opacification of the cortical layers and nucleus was observed after 48 and 72 h, respectively. Calcium chloride in our experiments also caused total lens opacification on days 3–4 (Figure 7b).

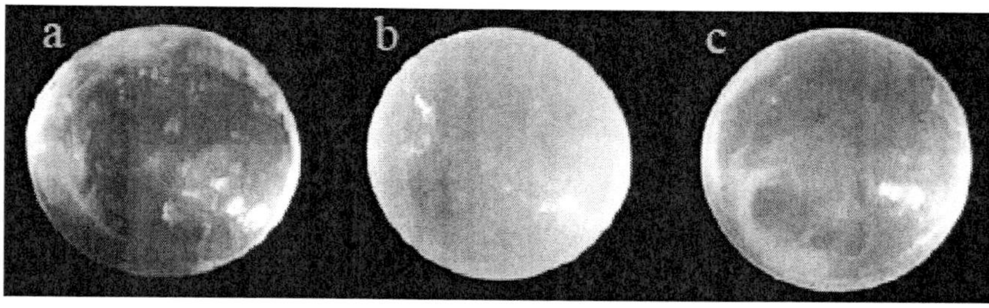

Figure 3. *Xenopus laevis* eye lenses cultured for four days in (a) the medium for amphibians without additives, (b) the same medium with 0.5 mM H_2O_2, and (c) the same medium with 0.5 mM H_2O_2 and RP from the bovine lens (10^{-12} mg/ml).

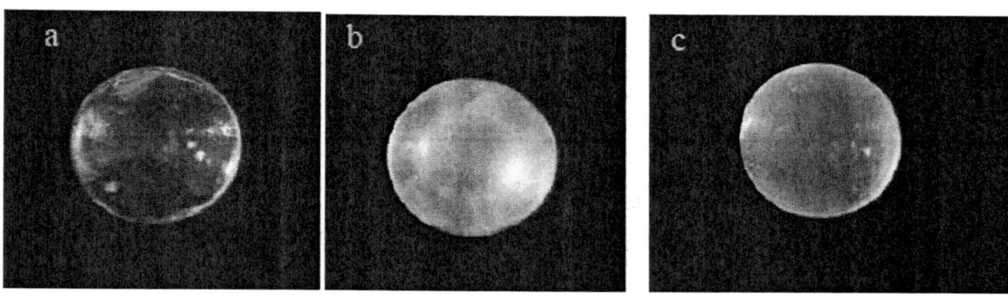

Figure 4. *Xenopus laevis* eye lenses cultured for four days in (a) the medium for amphibians without additives, (b) the same medium with 2.5 mM $CaCl_2$, and (c) the same medium with 2.5 mM $CaCl_2$ and RP from the bovine lens (10^{-12} mg/ml).

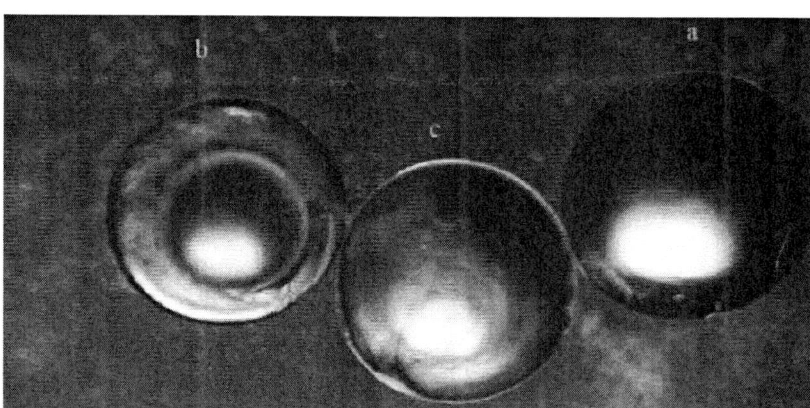

Figure 5. *Rana temporaria* eye lenses cultured for four days in (a) the medium for amphibians without additives, (b) the medium for amphibians containing 10 mM $CaCl_2$, and (c) the same medium containing 10 mM $CaCl_2$ and RP (10^{-12} mg/ml).

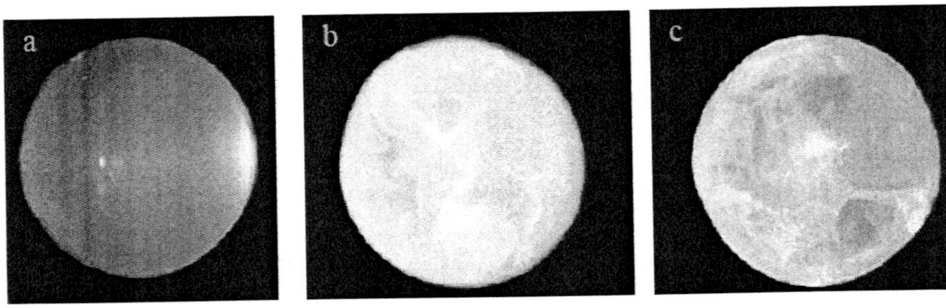

Figure 6. Bovine eye lenses cultured for three days in (a) the medium without additives, (b) the medium containing 20 mM CaCl2, and (c) the medium containing 20 mM CaCl2 and RP (10–12 mg/ml).

The effect of RP on cataract development in cultured rat lenses was studied using two RP preparations differing in the degree of purification: one preparation (RP1) was obtained immediately after IEF, and the other (RP2) was additionally purified by nondenaturing electrophoresis. Figures 7a and 7b show rat lenses cultured in the medium containing cataract-inducing agents together with RP1 and RP2, respectively. It is noteworthy that the preparation purified to a higher degree (RP2) was more effective in preserving lens transparency.

The results of visual assessment of lens opacification in different experimental variants fully coincided with the results of optical density measurements after the end of culturing (Figure 8). They show that the optical density of lenses cultured in the presence of RP was only half that of lenses exposed to cataract-inducing agents alone. Moreover, the protective effect of RP was manifested in the fact that lens transparency in the experiment was 1.5–2 times higher than in the lenses exposed to cataract-inducing agents alone, and its strength obviously depended on the degree of RP purification.

Spectrophotometric data (Figure 8) show that the rat eye lens treated with cataract-inducing agents, compared to the control, transmits much smaller amounts of visible light at different wavelengths, with the difference being approximately fivefold in the case of hydrogen peroxide and fourfold in the case of calcium chloride. Notably, this fact agrees with data that the total calcium content in human lenses with cortical cataracts is four times higher than in normal lenses [Tang et al., 2003]. According to these authors, high total calcium correlates with an increase in free intracellular calcium, which is due to reduced activity of Ca^{2+}-ATPase systems. The abundance of free intracellular calcium is favorable for the formation of calcium oxalate crystals, which, in turn, increase lens opacity [Tang et al., 2003].

Differences between hydrogen peroxide concentrations that are optimal for inducing cataract development in amphibian and mammalian lenses may be explained by differences in lens size. As this agent induces cataract by damaging the membranes of lens cells, its effective concentration must be proportional to the lens surface area. For example, the radius of *R. temporaria* lens (≈2mm) is approximately two times greater than the radius of *X. laevis* lens (≈1mm); and the radii of bovine (≈8.5mm) and rat (≈2mm) lenses differ by a factor of no less than four. On the other hand, the lenses of *R. temporaria* and rat are of approximately equal size, but hydrogen peroxide concentrations necessary for cataract induction are different: 2.0 and 0.5 mM, respectively. This may be due to species-specific differences in the state of important NADH- and NADPH-dependent enzyme systems of the eye lens [Rao and

Zigleer, 1992]. According to these authors, for example, the level of pyrimidine nucleotides in *R. temporaria* lens is markedly higher than in rat and *X. laevis* lenses.

Figure 7. Eye lenses of Wistar rats in experiments on cataract induction by (a) H_2O_2 and (b) $CaCl_2$ in the presence of different RP preparations: (*1*) medium 199 without additives, (*2*) medium 199 with a cataract-inducing agent alone, (*3*) medium 199 with a cataract-inducing agent and RP1 solution, and (*4*) medium 199 with a cataract-inducing agent and RP2 solution.

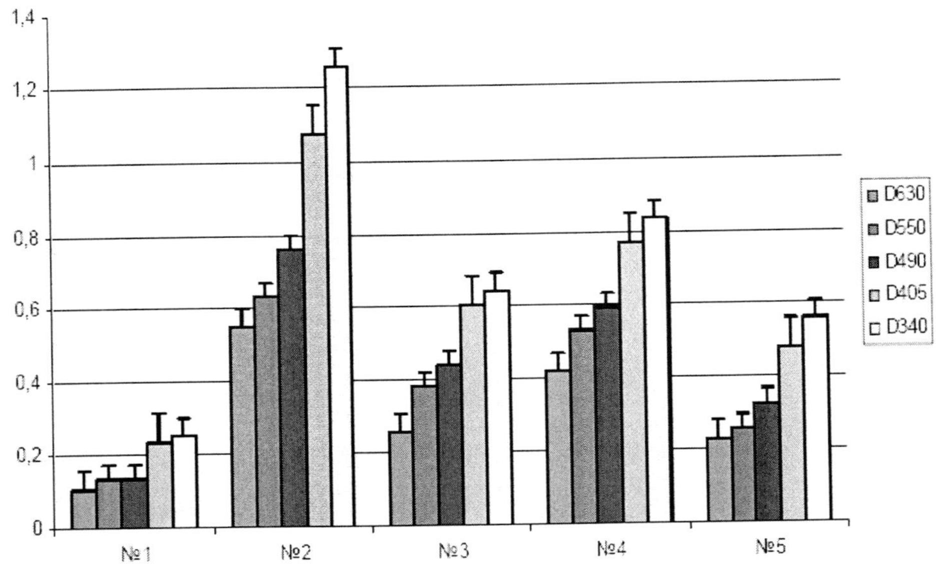

Figure 8. Spectrophotometric data on the effect of RP2 on cataract development induced by hydrogen peroxide and calcium chloride in the rat eye lens. The small bar on the right shows wavelengths at which measurements were made (nm) in different colors. The numbers of experimental series are shown on the horizontal axis: (1) medium 199, (2) medium 199 with H_2O_2, (3) medium 199 with H_2O_2 and RP2, (4) medium 199 with $CaCl_2$, (5) medium 199 with $CaCl_2$ and RP2. Ordinate shows optical density (A).

Likewise, minor differences between calcium chloride concentrations necessary for *in vitro* cataract induction in the lenses of different species are attributable to differences in the state of calcium-dependent enzyme systems [David and Shearer, 1984].

Attention should be paid to differences in the spatial pattern of lens opacification in amphibians, which were absent in mammals: hydrogen peroxide caused the development of a cortical cataract, whereas calcium chloride induced a nuclear cataract. In our opinion, this is additional evidence for differences in the activity of anticataract enzyme systems of the eye lens between animals representing different classes of vertebrates [David and Shearer, 1984; Rao and Zigleer, 1992; Siew and Bettelheim 1996; Tang et al., 2003].

Thus, in the context of published data, the results of this study may be regarded as evidence for differences in the regulation of calcium-dependent enzyme systems and the state of systems controlling lipid peroxidation in lens cell membranes between animal species used in our experiments. The lens-derived RP in ultralow doses can prevent the development of cataract, irrespective of the action mechanism of cataract-inducing agents, and its biological action is not species-specific.

The results of experiments on the *in vivo* development of traumatic cataract in rats were as follows. In the first series (after the eye was punctured to a depth of 5.1 mm), rats of the control group developed total traumatic cataracts (Figure 9). There were well-manifested consequences of strong posttraumatic inflammation, such as extended zones of cornea–lens capsule accretion and ingrowth of newly formed blood vessels, due to which the lens was difficult to isolate without damage. In the RP-treated group, the nuclear form of cataract was observed, with the cortical layers remaining transparent (Figure 9).

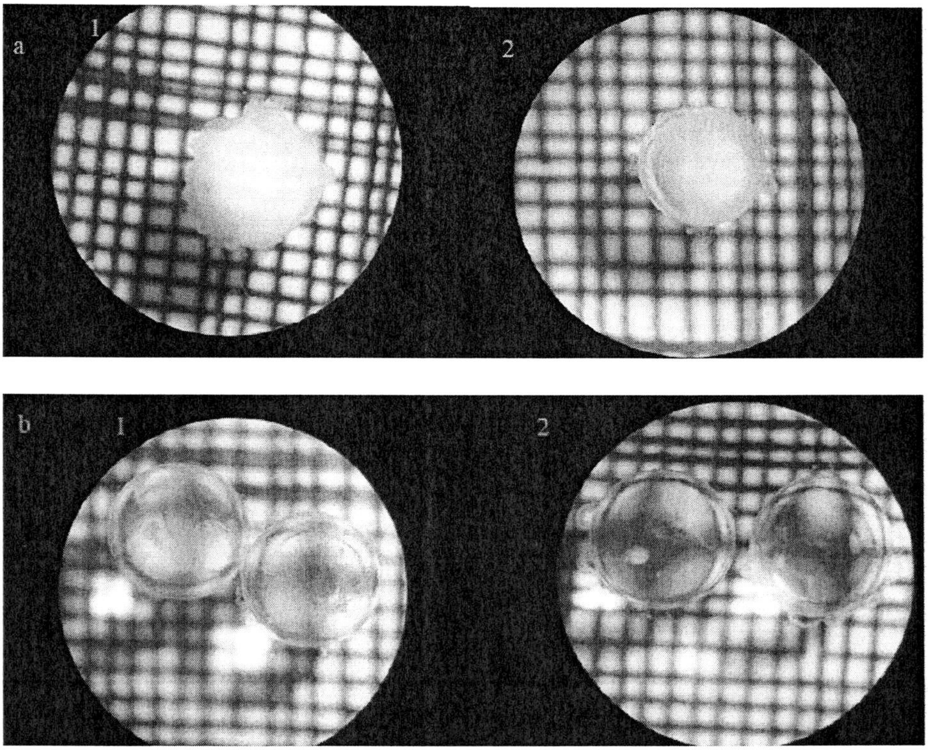

Figure 9. Eye lenses of rats used in experiments on traumatic cataract development *in vivo* (a) seven days after puncture to a depth of 5.1 mm and (b) seven days after puncture to a depth of 3.4 mm: (1) control, (2) experiment.

Inflammation was insignificant, and lenses were isolated undamaged. In the second series (after the eye was punctured to a depth of 3.4 mm), traumatic cataract was developed as an epithelial plug at the site where the lens capsule was punctured (Figure 10). This was observed in both control and RP-treated groups. In the latter group, however, the plug had a smaller area, lens epithelium proliferated more actively.

Thus, ultra low doses of the RP from the bovine eye lens can prevent the development of cataract both *in vivo* and *in vitro.* Generalizing the data presented above, we may conclude that this RP has a significant protective effect on this organ of other vertebrates, including both mammals and amphibians, under conditions leading to the development of cataract, irrespective of its etiology. It may well be that this factor triggers the regulatory systems that eventually restore the structural integrity of the lens, thereby providing for its transparency. As the RP is located in the lens epithelium, the protective effect of this factor may be due to its involvement in the regulation of main biological processes taking place in the lens capsule. It is well known that lens epithelium plays a key role both in the metabolism and the maintenance of morphological structure of the lens as a whole [Kopaeva, 2002].

Acknowledgements

This work was supported by a grant of the President of the Russian Federation (project № MK-3839.2005.4).

References

A. Bhatnagar, N.H. Ansari, A. Zacarias, S.K. Srivastava, *Experimental Eye Research*, Vol. 57, N 4, pp. 385-391, 1993.

A. Bhatnagar, N.H. Ansari, L. Wang, P. Khanna, C. Wang, S.K. Srivastava, *Experimental Eye Research*, Vol. 61, N 3, pp. 303-310, 1995.

K.C. Bhuyan and D.K. Bhuyan, *Current Eye Research,* Vol. 3, N 1, pp. 67-81, 1984.

X.L. Cui and M.F. Lou, *Experimental Eye Research,* Vol. 57, N 2, pp. 157-67, 1993.

L.L. David and T.R. Shearer, *Investigative Ophthalmology and Visual Science,* Vol. 25, N 11, pp. 1275-1283, 1984.

E. Engvall, P. Perlman, *Journal of Immunochemistry,* Vol. 8, pp. 871-874, 1971.

N.V. Engelhardt "Immunokhimicheskii analiz (Immunochemical Analysis)", Moscow, *Meditcina,* 1968, 165–188 pp.

"Immunoassays: a practical approach", Ed. By J. P. Gosling, New York, Oxford University Press. 2000. pp. 19 - 36.

B.D. Hames "One-dimensional gel electrophoresis of proteins", in the book *"Gel Electrophoresis of Proteins. A Practical Approach",* Ed. by B.D. Hames and D. Rickwood, New York, Oxford University Press, 1990, 106 pp.

V.G. Kopaeva. "The Lens", pp. 245–268, in the book "Glaznye bolezni" (Eye Diseases), Ed. by V.G.Kopaeva, Moscow, *Meditsina*, 2002, 560 pp.

M.S. Krasnov, E.N. Grigoryan, V.P. Yamskova, et al. *Ontogenez,* Vol. 31, N 5, pp. 368–373, 2000.

M.S. Krasnov, E.N. Grigoryan, and V.P. Yamskova, *Izvestia Rossiyskoy Akademii Nauk, Seria Biologicheskaya,* N 1, pp. 22–36, 2003a.

M.S. Krasnov, E.N. Grigoryan, V.P. Yamskova, D.V. Boguslavsky, I.A. Yamskov, *Radiatzionnaya biologia i radioecologia* N 3, pp. 265-268, 2003b.

M.S. Krasnov, E.P. Gurmizov, R.A. Gundorova, V.P. Yamskova, Yu.A. Kapitonov, *Ophtal'mologia,* Vol. 2, N 3, pp. 43-49, 2005.

U.K. Laemmli, *Nature*, vol. 227, pp. 680–685, 1970.

M.F. Lou, G.T. Xu, X.L. Cui, *Current Eye Research,* Vol. 14, N 10, pp. 951-958, 1995.

D.V. Margasyuk, M.S. Krasnov, V.P. Yamskova, E.N. Grigoryan, and I.A. Yamskov, *Oftal'mologiya,* Vol. 2, N 3, pp. 81–87, 2005b.

C.M. Rao and J.S.Jr. Zigleer, *Photochemistry and Photobiology,* Vol. 56, N 4, pp. 523-528, 1992.

E.L. Siew and F.A. Bettelheim, *Experimental Eye Research*, Vol. 62, N 3, pp. 265-270, 1996.

P.K. Smith, R.I. Krohn, G.T. Hermanson, A.K. Mallia, F.H. Gartner, M.D. Provenzano, E.K. Fujimoto, N.M. Goeke, B.J. Olson, D.C. Klenk, *Analitical Biochemistry*, Vol. 150, N 1, pp. 76-85, 1985.

D. Tang, D. Borchman, M.C. Yappert, G.F.J.M. Vrensen, V. Rast, *Investigative Ophthalmology and Visual Science,* Vol. 44, N 5, pp. 2059-2066, 2003.

S.Yu Venyaminov, J.T. Yang "Determination of protein secondary structure" In the book *"Circular Dichroismand the conformational analysis of Biomolecules",* Ed by G.D.Fasman, New York, Plenum Press, pp. 69-107, 1996.

I.A. Yamskov, V.P. Yamskova, A.N. Danilenko, Z.S. Klemenkova, B.G. Antipov, F.R. Chernikov, M.M. Gusynina, and E.Yu. Rybakova, *Rossiyskii Khimicheskii Zhurnal* (Zh. Ross. Khim. O–va im. D.I. Mendeleeva), Vol. 43, N 5, pp. 34–39, 1999.

V.P. Yamskova, E.A. Modyanova, M.M. Reznikova, and A.G. Malenkov, *Molecularnaya Biologia,* Vol. 11, N 5, pp. 1147–1154, 1977.

V.P. Yamskova and M.M. Reznikova, *Zhurnal Obshchey Biologii,* Vol. 45, N 3, pp. 373–382, 1984.

V.P. Yamskova and M.M. Reznikova, *Zhurnal Obshchey Biologii,* Vol. 52, N 2, pp. 181–191, 1991.

In: Biochemical Physics Frontal Research
Editors: Sergei D. Varfolomeev et al., pp. 35-45
ISBN 1-60021-425-8

Chapter 5

REGULATORY PROTEINS FROM THE MAMMALIAN LIVER THAT DISPLAY BIOLOGICAL ACTIVITY AT ULTRA LOW DOSES

A. V. Borisenko, V. P. Yamskova, M. S. Krasnov, I. V. Blagodatskikh*, V. V. Vecherkin* and I. A. Yamskov*

Kol'tsov Institute of Developmental Biology, Russian Academy of Sciences,
26 Vavilova str, Moscow, 11991, Russia
*Nesmeyanov Institute of Organoelement Compounds,
Russian Academy of Sciences, 28 Vavilova str, Moscow, 119991, Russia

ABSTRACT

A protein extract from the rat liver was analyzed for the presence of regulatory proteins (RPs) displaying biological activity at ultra low doses. Four fractions containing such proteins were identified, and their physicochemical properties were studied. To assess the specific activity of RPs, an original in vitro model of roller organ culture of the newt liver was developed. Using this model, the acidic hepatic regulatory protein was shown to have a protective effect on the liver.

Keywords: Liver, regulatory proteins, ultra low doses, roller organ culture

In our previous studies, an adhesion protein that proved to be biologically active at ultra low doses (10^{-8}–10^{-12} mg total protein/ml), producing an effect on the properties of the plasma membrane, adhesion, and proliferation of hepatocytes, was detected in the mammalian liver [Yamskova et al., 1977a, 1977b; Malenkov et al., 1978]. Proteins displaying biological activity at ultra low doses were subsequently revealed in different animal tissues, including the glandular tissue and blood serum [Yamskova and Reznikova, 1991; Yamskova et al., 1994; Yamskov et al., 1999]. Their physicochemical properties were similar: these were thermostable low-molecular glycosylated proteins with a specific secondary structure, molten globule-like molecular conformation, and a strong tendency to form large intermolecular aggregations. In tissue extracts obtained under special conditions, the proteins of interest were represented by several acidic fractions (pI < 3.0 and pI 4.5–5.1) and basic fractions (pI > 8.5) [Yamskova and Reznikova, 1991; Yamskov et al., 1999; Krasnov et al., 2003b]. Acting at ultra low doses, these proteins produced an effect on cell adhesion, migration, proliferation, and differentiation [Yamskova and Reznikova, 1991; Yamskov et al., 1999, 2001; Krasnov et al., 2003b] and stimulated regeneration and repair of pathologically altered tissues [Gundorova et al., 1997]. In the aggregate, the above data provided evidence for the involvement of these proteins in the regulation of homeostasis in vertebrates, and, hence, they were named regulatory proteins (RPs).

The purpose of this study was to study RPs from the mammalian liver. RP isolation and subsequent purification followed the procedure previously developed for this group of proteins. To study specific effects of hepatic RPs, an original *in vitro* model of roller organ culture of the newt liver was developed.

EXPERIMENTAL

Wistar rats (both sexes, body weight 210-230 g), male F1 hybrid mice (C57BL/CBA) (20-22 g), and *Pleurodeles waltl* newts were from the vivarium (aquarium) of the Kol'tsov Institute of Developmental Biology. All reagents were of chemical purity grade. All solutions were made in deionized water (15 MΩ) from a MilliQ column.

Rats were anesthetized with diethil ether and decapitated. The liver was perfused through the portal vein with a cold (4–8°C) buffered saline (0.15M NaCl, 1mM $CaCl_2$, 5 mM KCl, 1 mM HEPES; pH 7,2; 10ml; 1min), cut into tissue fragments (1-2 g) and placed in the same saline. The liver tissue collected from 50 rats (430 g fresh weight) was extracted with 250 ml of the saline for 2 h. The resulting extract was decanted and centrifuged at 3000 *g* for 10 min to remove blood cells. All these manipulations were performed at 4–8°C. Subsequent steps followed the procedure developed in our previous studies [Yamskova and Reznikova, 1991], which involved precipitation of protein with ammonium sulfate from the extract (repeated twice) and its fractionation by isoelectric focusing (IEF) in a sucrose density gradient, with spectrophotometric detection of protein at 280 nm.

Fractions collected for analysis were as follows: (1) acidic proteins with pI < 3.0, (2) proteins with pI 4.5–5.1, (3) proteins with pI 6.7–7.2, and (4) basic proteins with pI 8.5–9.5.

Fractions of rat liver extract were analyzed for the presence of RPs using a cell adhesion test developed for biological indication of proteins belonging to this group [Yamskova and Reznikova, 1991].

Sucrose and ampholytes were removed by dialysis against deionized water (fractions 1, 2, and 4) or the saline used for extraction (fraction 3), and the protein composition of the fractions was analyzed by SDS-electrophoresis in 12% PAAG [Laemmli, 1970] in a Mini Protean-3 cell (BioRad, Italy). A molecular weight marker kit for SDS gel electrophoresis (LKB, Sweden) was used as reference.

The fractions obtained after IEF were further separated by nondenaturing electrophoresis in 12.5% PAAG [Hames, 1990]. The band corresponding to proteins with Rf = 0.90 ± 0.05 was cut out and incubated in 1 ml of water to elute these proteins (3 times at 4–8°C). The eluates were pooled, dialyzed against distilled water, and lyophilized. The resultant preparation was analyzed by physicochemical methods and tested on the *in vitro* model of roller organ culture of the newt liver.

To assess the secondary structure of test proteins, their circular dichroism spectra were recorded in a Jasco 720 dichrograph (Japan) equipped with a computer data processing system. Their analysis was performed using the k2d program [Venyaminov, Yang, 1996].

The hydrodynamic radius of protein particles was determined by measuring dynamic light scattering with a PhotoCor Complex device (PhotoCor, Russia) equipped with an automatic goniometer and PhotoCor-FC real-time correlator. A 20-mW Uniphase 1135P helium–neon laser was used as a light source. Protein solutions were filtered through Durapore membranes (Millipore) with pore size 0.45 μm to remove dust, and measurements were made at a monochromatic laser radiation wavelength of 633 nm and a scattering angle of 90°, at 25°C. The homodyne correlation function of intensity $G^{(2)}(t)$ [Stepanek, 1993] was determined in a delay time range of 2×10^{-8} to 10^3 s. The distribution by correlation times was calculated by means of the inverse Laplace transform using the CONTIN program [Provencher, 1985]. The hydrodynamic radius (R_h) was determined from data on correlation time (τ) using the equations $D=1/q^2\tau$ and $D=kT/6\pi\eta R_h$, where η is solvent viscosity, κ is the Boltzmann constant, D is diffusion coefficient, and q is the wave vector.

Tests on the roller organ culture of the newt liver were carried out as follows. Newts were anesthetized with urethane, the liver was isolated, and its median lobe was cut into fragments (3 × 3 mm), which were placed into vials with the culture medium for amphibians: 64 ml of medium 199, 7.14 mg of HEPES, 9 ml of fetal calf serum, and water to a final volume of 100 ml. The medium was supplemented with 0.8 mg/ml gentamicin. The RP preparation was added to a final concentration of 10^{-12} mg protein/ml, and the vials were placed in an RM 5 roller (Assistant, Germany) adjusted to 35 rpm and cultivated at 21–23°C for three days. The cultures were then fixed in the Bouin solution, dehydrated, and embedded in paraffin. Histological section (3–4 μm) were stained with hematoxylin–eosin and analyzed under a microscope to measure the area of pigmented cell clusters.

Antisera for immunohistochemical identification of RPs were obtained by immunizing rabbits with preparations of acidic RPs from the rat liver according to the conventional procedure [Gosling, 2000]. The antigen solution was mixed with the equal volume of complete Freund's adjuvant injected into male Chinchilla rabbits weighing approximately 1 kg. Each rabbit received 1 ml of the mixture injected under the skin on the back and into the femoral muscles at approximately equal doses. Thereafter, injections of the antigen mixed with incomplete Freund's adjuvant (1:1) were made in the same way at seven-day intervals. Blood samples taken from the marginal ear vein beginning from day 14 after the first injection, at the same intervals, were tested for the presence of specific antibodies by means of ELISA [Engvall and Perlman, 1971].

The localization and expression of acidic RPs from the rat liver in the mouse liver were studied by an indirect immunofluorescent method [Engelhardt, 1968] with FITC-labeled secondary antibodies to rabbit IgG. The liver was fixed in 4% formaldehyde solution in 0.01 M phosphate-buffered saline (PBS, pH 7.4) for 6 h, washed with PBS for 24 h at 4°C, dehydrated, and embedded in paraffin. Transverse histological sections (3–4 μm) were mounted on gelatin-coated glass slides. Before incubation with antibodies, the sections were deparaffined, washed with PBS, and incubated in 1% solution of bovine serum albumin (Sigma, United States) in PBS to prevent unspecific antibody binding. Thereafter, the sections were washed in fresh PBS (3 × 5 min), placed in a humid chamber, and incubated overnight with the antiserum against the protein fraction of interest (dilution 1:10) at room temperature. After repeated washing with PBS (3 × 5 min), the sections were treated with the secondary, FITC-conjugated antibodies to rabbit IgG (whole molecule) (Sigma, United States) diluted 1:35. The secondary antibodies were washed out, and the sections were embedded in glycerol with PBS (9:1) and protected with coverslips. Control sections were treated with (1) secondary antibodies alone or (2) nonimmune rabbit serum. Immunofluorescence was analyzed and recorded using equipment manufactured under an ABC microscope (Olympus, Japan) equipped with a DEF digital camera (Leica, Germany).

Experimental data were processed by methods of variation statistics, with Student's *t*-test used to estimate significance of differences.

RESULTS AND DISCUSSION

Figure 1 shows the pattern of fractions after IEF of the rat liver extract. With regard to the results of our previous studies [Yamskova et al., 1977, 1978; Yamskova and Reznikova, 1991; Yamskov et al., 1999], fractions collected for analysis were as follows: fraction 1, acidic proteins with pI < 3.0; fraction 2, which corresponded to the fraction of RP previously detected in mammalian blood sera (pI 4.5–5.1); fraction 3, which corresponded to the fraction of RP previously detected in the liver (pI 6.7–7.2); and fraction 4, basic proteins with pI 8.5–9.5.

In biological tests, these fractions displayed biological activity when tested at ultra low doses, which indicated that they did contain RPs (Table 1). As an example, consider the diagram showing the membranotropic effect of the acidic hepatic RP (initial protein concentration 0.1 mg/ml) (Figure 2). The data confirming the presence of acidic and basic RP fractions in the liver extract agree with the results of studies on other RPs identified in different animal and plant tissues [Yamskova and Reznikova, 1999; Yamskov et al., 1999, 2001; Krasnov et al., 2003a, 2003b).

In the course of purification, fractions 1, 2, and 4 proved to be resistant to the influences of temperature, low ionic strength (deionized water), and pH of the medium. Moreover, they retained biological activity even after repeated freezing (–55 to –60°C), thawing, and lyophilization. Such a resistance to physicochemical factors indicates that RPs contained in the test fractions are similar in their properties to RPs isolated by the same methods from other tissues [Yamskov et al., 2003; Krasnov et al., 2003a, b]. As to fraction 3, its dialysis against deionized water resulted in precipitation of a white flaky substance, with the following loss of biological activity. Hence, this fraction was dialyzed against the saline used for extraction. This is evidence that fraction 3 may be identical to the hepatic RP identified in our previous studies, which had its isoelectric point in the same range (pI 6.7–7.2) and was also sensitive to physicochemical factors [Yamskova et al., 1977a; Yamskova, 1978; Yamskova et al., 1978]. This RP was first detected in the liver extract obtained under different conditions: at 25°C and using Ca^{+2}-free solutions.

The fact that fraction 2 of the liver extract displayed biological activity at ultra low doses indicates that the corresponding RP (pI 4.5–5.1) may be identical to the RP identified in vertebrate blood sera [Yamskova and Reznikova, 1991; Yamskova et al., 2001]. This hypothesis agrees with published data that the bulk of blood serum proteins is synthesized in the liver [Engelhardt, 1984]. To confirm it, we conducted comparative biological tests of fractions collected in the pH range 4.5–5.1 after IEF of crude extracts from the rat liver and lungs and supernatate fractions obtained after precipitation of protein with ammonium sulfate from the crude extracts. The results showed that fractions isolated from the liver extracts, acting at ultra low doses, had a membranotropic effect *in vitro* on both liver and lung cells; i.e., their biological action was not tissue-specific. It is noteworthy that RP identified in vertebrate blood sera had a similar unspecific membranotropic effect [Yamskova and Reznikova, 1991]. Corresponding fractions of the lung extracts lacked biological activity. Therefore, the results of these experiments provide evidence for the hypothesis that the RP found in blood serum is synthesized in the liver.

Table 1. Characteristics of hepatic RPs determined by isoelectric focusing

Fraction	pI range	Total protein, mg/ml	Volume, ml	Dilutions at which biological activity is manifested
1	< 3.0;	0.01	26.5	10^{12}-10^{14}
2	4.5–5.1;	0.015	34.3	10^{11}-10^{15}
3	6.7–7.2;	0.005	28.0	10^{4}-10^{5}
4	8.5–9.5	0.09	24.0	10^{10}-10^{11}

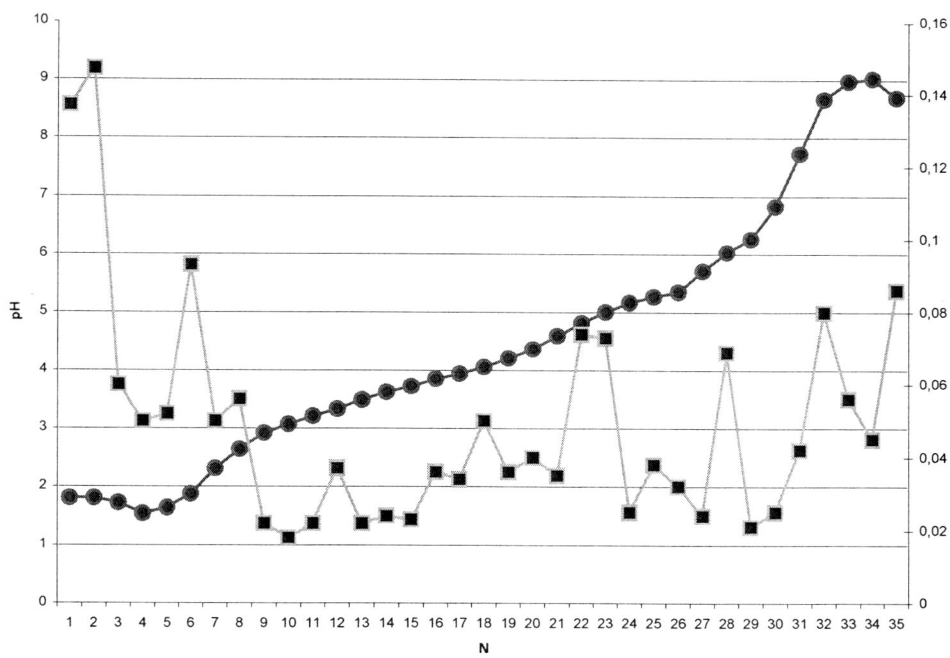

Figure 1. Pattern of fractions after isoelectric focusing of extract from the rat liver. Ordinates 1 and 2 shows pH and A_{280}, respectively; abscissa shows number of fractions.

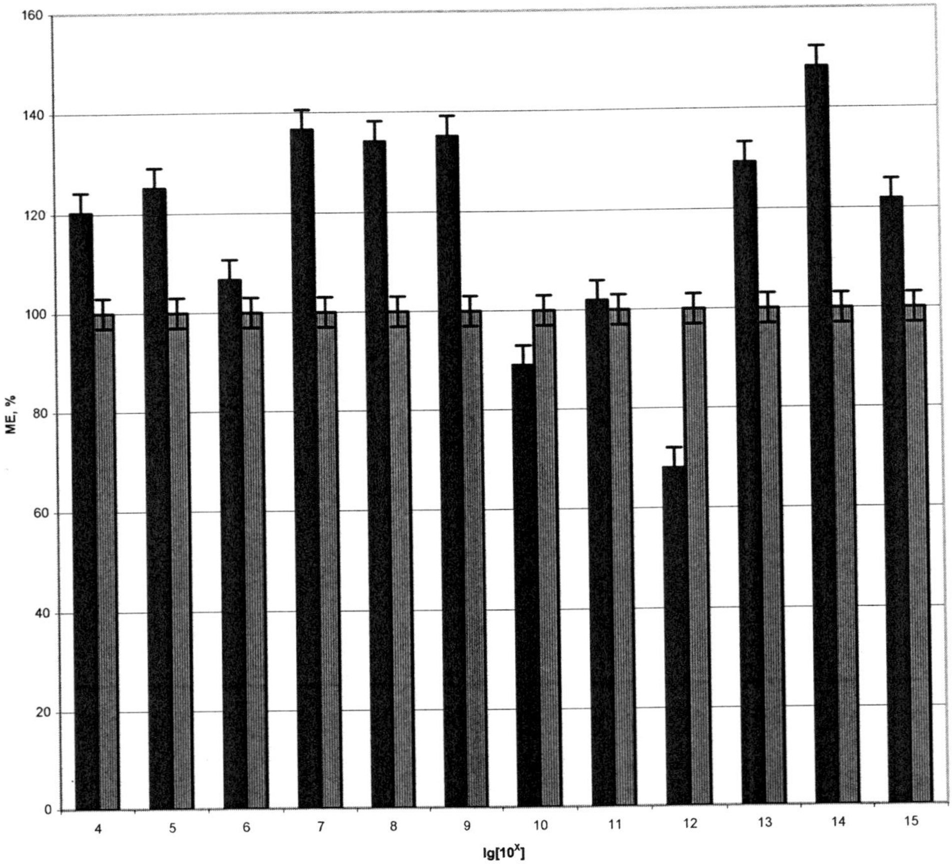

Figure 2. Membranotropic effect of acidic hepatic RP in biological tests (black bars) versus control (white bars). Ordinate shows membranotropic effect (ME, %) [Yamskova and Reznikova, 1999]; abscissa shows dilution factors of RP preparation (lg[10^x]), initial concentration 0.1 mg/ml.

Figure 3 shows the protein pattern of RP fractions from the rat liver after electrophoresis under denaturing conditions. It is seen that the main component of each fraction is a low-molecular-weight protein (< 10 kDa). We electrophoretically separated fractions 1, 2, and 4 in 12% PAAG under nondenaturing conditions and collected the corresponding proteins. Biological tests showed that all three preparations at ultra low doses had a membranotropic activity.

The secondary structure of the three low-molecular-weight RPs from the liver was analyzed by their circular dichroism spectra. The results showed that the main structural elements were antiparallel and parallel β-sheets and folds, with the remainder of the molecule being a stochastic globule. Table 2 shows the percent ratio of these elements in the molecule of the acidic hepatic RP. The structural parameters of hepatic RPs analyzed in this study are closely similar to those of corresponding RPs isolated from other mammalian tissues [Yamskov et al., 2001].

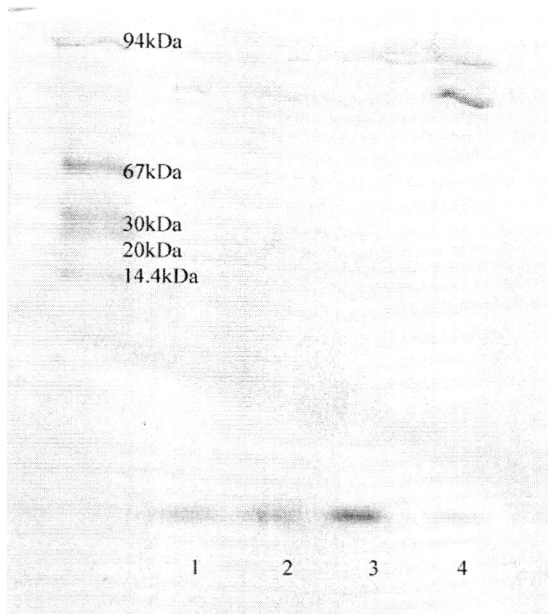

Figure 3. SDS-electrophoresis of RP fractions from the liver in 12% PAAG: (*1*) fraction 1 (pH < 3.0), (*2*) fraction 2 (pI 4.5–5.1), (*3*) fraction 3 (pI 6.7–7.2), and (*4*) fraction 4 (pI 8.5–10.0).

Table 2. Parameters of secondary structure of the acidic hepatic RP

Elements of secondary structure	Content, %
α-Helix	2.2±0.5%
Antiparallel β–sheet	48.1 ± 0.5%
Parallel β–sheet	3.7 ± 0.5%
β-Fold	16.7 ± 0.5%
Random coil	29.3 ± 0.5%
Total	100.0 ± 0.5%

As follows from published data, the prevalence of β-sheet and random in the protein folding pattern imply that these proteins should have a well-manifested tendency to form intermolecular aggregations [Ross and Poirier, 2004]. As shown by the method of dynamic light scattering, the hydrodynamic radius of protein particles averaged 40–60 nm, and the curves of their distribution by size were unimodal. As an example, Figure 4 shows the size distribution of particles for fraction 1.

To assess the specific biological activity of RPs from the rat liver, we developed the *in vitro* model of roller organ culture of the newt liver. Its advantage is that newt tissues, unlike mammalian tissues, show no significant destructive changes in long-term organ culture. This model proved to be applicable to our study because the biological effect of RPs is not species-specific [Krasnov et al., 2000].

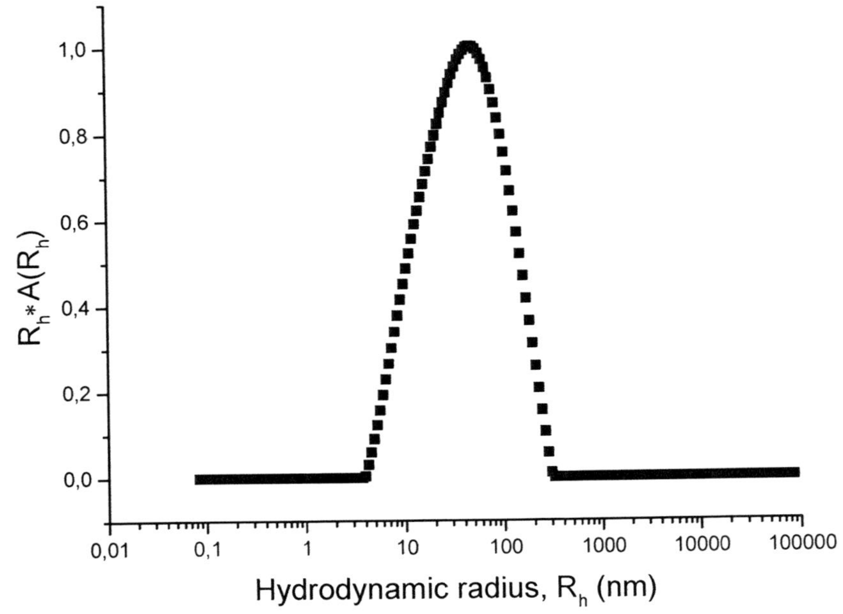

Figure 4. Distribution of particles of the acidic hepatic RP by hydrodinamic radius.

Figure 5 shows a histological section of the normal newt liver, in which hematopoietic and pigmented cells can be distinguished, in addition to hepatocytes. Pigmented cells are melanomacrophages, analogs of Kupffer cells of the mammalian liver [Sichel et al., 2002]. They phagocytize various particles larger than 10 nm in size, such as microorganisms, viruses, and cell debris, and account for detoxification of various xenobiotics [Frangioni et al, 2005]. Thus, liver melanomacrophages play a major role in the protective systems of newts.

Pilot experiments with this model showed that the viability of liver cultures strongly depended on temperature. The optimal temperature of the medium was 20 ± 1°C; at 25 ± 1°C, all cells perished within 24 h; at 23 ± 1°C, part of hepatocytes entered apoptosis, and an increase in the number of pigmented cells was observed.

In this study, we assessed protective properties of the acidic RP from the rat liver (10^{-12} mg protein/ml) using organ cultures of the newt liver maintained at 23 ± 1°C. On day 3, an analysis of the cultures treated with RP revealed a statistically significant increase in the number of pigmented cells, which became approximately twice greater than in the control cultures (Figure 6). Therefore, this RP can exert influence on melanomacrophages of the amphibian liver, stimulating their proliferative activity or migration into tissues. It should be noted that immunochemical localization of this RP revealed its presence in the sinusoid of the mouse liver (Figure 7), which indicates that the acidic hepatic RP in mammals is synthesized by cells of the sinusoid (probably, by Kupffer cells).

In view of the concept concerning specific functional features of pigmented cells in the amphibian liver [Frangioni et al, 2005], the results of our study provide evidence that the RP isolated from the rat liver has well-manifested hepatoprotective properties. This agrees with previous data that very low concentrations of hepatic acidic RP stimulated the functional activity of cytochrome P-450 in rats [Kotelevtsev et al., 2000]

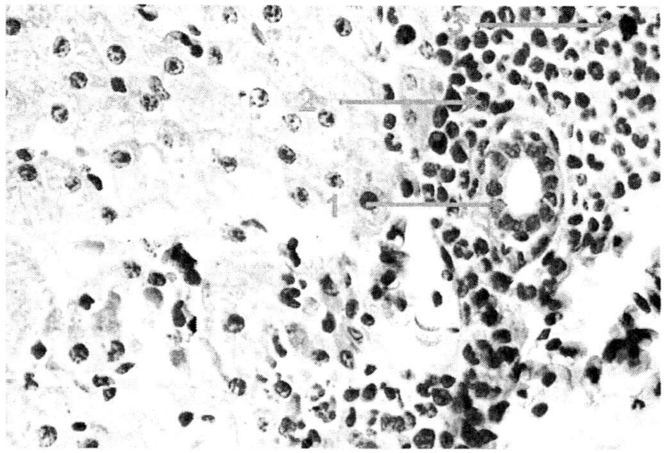

Figure 5. Histological section of normal newt liver: (1) duct, (2) zone of hematopoiesis, (3) single pigmented cells. Magnification: ocular 10×, objective lens 20×.

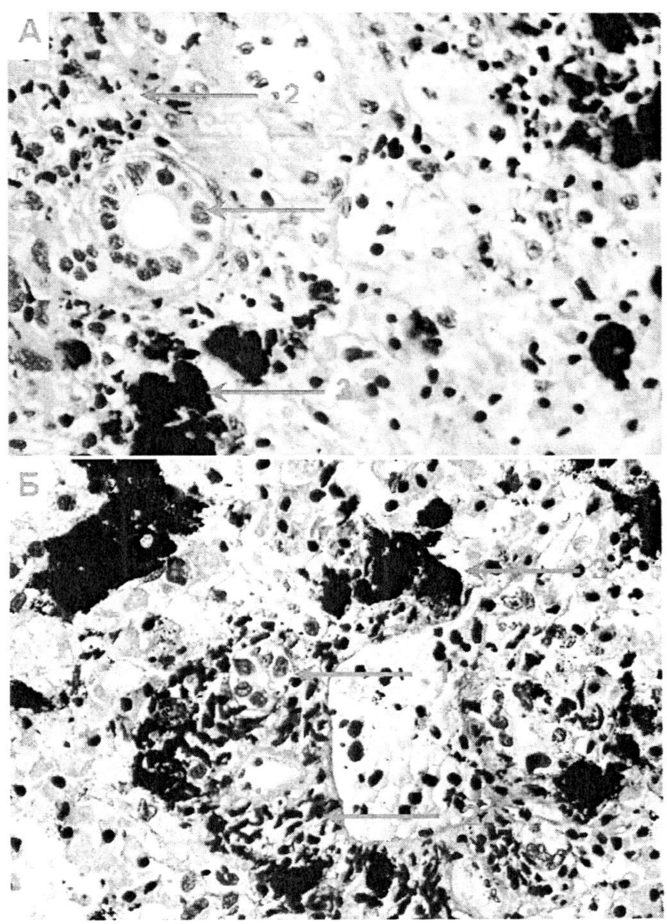

Figure 6. Control (a) and RP-treated (B) cultures of the newt liver on day 3: (1) duct, (2) zone of hematopoiesis, (3) pigmented cell clusters. Magnification: ocular 10×, objective lens 20×.

Figure 7. Immunochemical localization of acidic hepatic RP in the sinusoid of the mouse liver (indicated with an arrow). Magnification: ocular 10×, objective lens 40×.

Thus, this study confirmed that the liver contains RPs displaying biological activity at very low concentrations. Acidic and basic fractions of hepatic RPs were identified, and data were obtained that the liver is a probable source of RPs found in blood sera of vertebrates. Using an original model of roller organ culture of the newt liver, a protective effect of hepatic RPs on the liver was demonstrated.

References

N. V. Engelhardt, Immunochemical Analysis, Moscow, *Meditcina,* pp 165–188, 1968.

N. V. Engelhardt "Synthesis of blood plasma proteins and alpha-fetoprotein in ontogeny", pp. 99-100, in the book "Immunologicheskie aspekty biologii razvitiya", Moscow, *Nauka,* 1984, 346 pp.

E. Engvall, *Immunochemistry*, Vol. 8, N 9, pp. 871–874, 1971.

G. Frangioni, M. Santoni, S. Bianchi, M. Franchi, G. Fuzzi, S. Marcaccini, C. Varlani, G.Borgioli, *Journal Experimental Zoology*, Vol. 303A, pp. 123–131, 2005.

J. P. Gosling, "Immunoassays", pp. 19-36, *"A Practical Approach",* Ed by J.P. Gosling, New York, Oxford University Press, 2000, 215 pp.

R. A. Gundorova, I.P. Khoroshilova-Maslova, E.V. Chentsova, L.V. Ilatovskaya, V.P. Yamskova, I.Yu. Romanova, *Vestnik Oftal'mologii,* Vol. 113, N 2, pp. 12–15, 1997.

B. D. Hames "One-dimensional gel electrophoresis of proteins", pp. 1-44, in the book *"Gel Electrophoresis of Proteins. A Practical Approach"*, Ed. by B.D. Hames, D. Rickwood, New York, Oxford University Press, 1990, 106 pp.

S. V. Kotelevtsev, V.P. Yamskova, I.A. Yamskov, F.F. Nagdaliev, V. Truare, *Ontogenez,* Vol. 31, N 4, pp. 274–276, 2000.

M. S. Krasnov, V.P. Yamskova, *Ontogenez,* Vol. 31, N 4, pp. 281–282, 2000.

M. S. Krasnov, D.V. Margasyuk, I.A. Yamskov, V.P. Yamskova, Radiatsionnaya. *Biologiya Radioekologiya,* N3, pp. 269–272, 2003a.

M. S. Krasnov, E.N. Grigoryan, V.P. Yamskova, Izvestiya Rossiiskoi Akademii Nauk, *Seriya Biologicheskaya,* N 1, pp. 22–36, 2003b.

U. K. Laemmli, *Nature,* Vol. 227, pp. 680–685, 1970.

A. G. Malenkov, E.A. Modyanova, V.P. Yamskova, *Tsitologiya,* Vol. 20, N 8, pp. 957–962, 1978.

S. W. Provencher, *Macromolecula Chemistry,* Vol. 82, N 15, pp. 632, 1985.

C. A. Ross, M.A. Poirier, *Nat. Med.*, Vol. 10, Suppl., pp. 10–17, 2004.

G. Sichel, M. Scalina, C. Corcoro, *Microsc. Res. Techn.*, Vol. 57, pp. 477–490, 2002.

P. Stepanek *"Dynamic Light Scattering. The Method and Some Applications."* Ed. by W. Brown, Oxford, Clarendron Press, 1993, 177 pp.

I. A. Yamskov, V.P. Yamskova, A.N. Danilenko, Z.S. Klemenkova, B.G. Antipov, F.R. Chernikov, M.M. Gusynina, E.Yu. Rybakova, *Rossiiskii Khimicheskii Zhurnal,* Vol. 43, N 5, pp. 34–39, 1999.

I. A. Yamskov, A.A. Vinogradov, A.N. Danilenko, L.A. Maslova, V.P. Yamskova *Prikladnaya Biokhimiya Mikrobiologiya*, Vol. 37, N 1, pp. 36–42, 2001.

V. P. Yamskova, *Biofizika*, Vol. 23, pp. 428–432, 1978.

V. P. Yamskova, M.M. Reznikova, *Zhurnal Obshei Biologii,* Vol. 52, N 2, pp. 181–191, 1991.

V. P. Yamskova, E.A. Modyanova, V.I. Leventhal, T.P. Lankovskaya, O.K. Bocharova, A.G. Malenkov, *Biofizika,* Vol. 22, pp. 168–174, 1977a.

V. P. Yamskova, E.A. Modyanova, M.M. Reznikova, A.G. Malenkov, *Molekulyarnaya Biologiya,* Vol. 11, N 5, pp. 1147–1154, 1977b.

V. P. Yamskova, N.V. Nechaeva, N.B. Tumanova, Yu.G. Yurovitskii, T.E. Novikova, V.I. Fateeva, Izvestiya Rossiiskoi Akademii Nauk, *Seriya Biologicheskaya*, N 2, pp. 190–196, 1994.

25. S. Yu. Venyaminov, J.T. Yang "Determination of protein secondary structure. In *CircularDichroismand the conformational analysis of Biomolecules.",* Ed. by G.D. Fasman Plenum Press., New York, 1996, 69-107 pp.

In: Biochemical Physics Frontal Research
Editors: Sergei D. Varfolomeev et al., pp. 47-59
ISBN 1-60021-425-8

Chapter 6

REGULATORY PROTEIN FROM BOVINE CORNEA: LOCALIZATION AND BIOLOGICAL ACTIVITY

D. V. Margasyuk*[*]*, M. S. Krasnov*[**]*, I. V. Blagodatskikh*[*]*,
***E. N. Grigoryan*[**]*, V. P. Yamskova*[**] *and I. A. Yamskov*[*]**

[*]Nesmeyanov Institute of Organoelement Compounds,
Russian Academy of Sciences, ul. Vavilova 28, Moscow,
117813 Russia; fax (495)135-50-37
[**]Kol'tsov Institute of Developmental Biology,
Russian Academy of Sciences, ul. Vavilova 26, Moscow, 117334 Russia

ABSTRACT

A regulatory protein (RP) biologically active at ultra low doses was isolated from the bovine cornea and analyzed for physicochemical properties and activity in biological tests. The RP was localized in the intercellular space of corneal epithelium and endothelium in rats and newts, and its content was shown to increase with time in organ cultures of the cornea. This RP proved to stimulate cell proliferation and migration in the corneal epithelium and to aid in the maintenance of corneal tissue structure *in vitro*.

Keywords: Regulatory protein, ultralow concentrations, cornea, localization, proliferation, roller organ culture

INTRODUCTION

In the course of our previous studies, regulatory proteins (RPs) exhibiting biological activity at ultra low doses (10^{-8}–10^{-12} mg/ml) were identified in different tissues of vertebrates [Yamskova et al., 1977; Yamskova and Reznikova, 1991; Yamskov et al., 1999]. Extremely small doses of these RPs produced an effect on cell proliferation, differentiation, and adhesion, as well as on the level of protein synthesis and the activity of enzyme systems within the cells [Yamskova et al., 1977; Bueverova et al., 1985; Yamskova et al., 1994; Krasnov et al., 2000]. The results of these studies provided a basis for the concept of organ–tissue homeostasis [Yamskova and Yamskov, 1999].

With respect to physicochemical properties, RPs are characterized by a low molecular weight (< 6 kDa), secondary structure with the prevalence of β-sheet regions, and tendency to form high-molecular complexes [Yamskova et al., 2004]. The biological activity of RPs from animals tissues is tissue-specific but lacks species specificity [Krasnov et al., 2003; Margasyuk et al., 2005a]. The purpose of this study was to analyze the physicochemical properties and biological activity of RP isolated from the bovine cornea (CRP).

EXPERIMENTAL

Bovine eyes were obtained from the Taganka Meat-Packing Plant, Moscow. Wistar rats and *Pleurodeles waltl* newts were from the vivarium (aquarium) of the Kol'tsov Institute of Developmental Biology, Russian Academy of Sciences.

Corneas dissected from 20 bovine eyes were processed according to the procedure worked out in our previous studies, which involved extraction, ballast protein precipitation with ammonium sulfate, isoelectrofocusing (IEF) of remaining proteins in a sucrose gradient with ampholytes, and nondenaturing electrophoresis in 7.5% PAAG [Hames, 1990; Yamskova and Reznikova, 1991].

The apparent molecular weight of CRP was determined by SDS-electrophoresis in 12.5% PAAG [Laemmli, 1970]. Parameters of its secondary structure were inferred from circular dichroism spectra recorded in a Jasco 720 dichrograph (Japan) and processed using the k2d program [Venyaminov, Yang 1996].

The hydrodynamic radius of particles formed by CRP in solutions was estimated by measuring dynamic light scattering PhotoCor Complex device (PhotoCor, Russia) equipped with a PhotoCor-FC real-time correlator. A 20-mW Uniphase 1135P helium–neon laser was used as a light source. The solutions were purified of dust by filtering through Durapore membranes with a pore size of 0.45 μm (Millipore, United States), and measurements were made at a monochromatic laser radiation wavelength of 633 nm and a range of a scattering angle of 60° to 120°, at 25°C. The homodyne correlation function of intensity $G^{(2)}(t)$ [Stepanek, 1993] was determined in a delay time range of 2×10^{-3} to 10^{3} s. Distribution of correlation time was made by the method of inverse Laplace transform using the CONTIN [Provencher, 1985]. For transition from times of correlation (τ) to hydrodynamical radius (R_h) used following parities $D=1/q^2\tau$ and $D=kT/6\pi\eta R_h$, where η - вякость solvent, k – Bolzmann's constant, D - factor of diffusion, q - a wave vector.

The biological activity of CRP-containing fractions and its dependence on concentration were assessed in the cell adhesion tests developed for biological indication of RPs [Yamskova and Reznikova, 1991]. Protein concentration in the fractions was determined by the colorimetric method with bicinchoninic acid [Smith et al., 1985].

Rabbit antisera to CRP were obtained by conventional methods [Gosling, 2000; Margasyuk et al., 2005b]. They were used for immunohistochemical localization of CRP and estimation of its content in the cornea by an indirect immunofluorescent method [Engelhardt,

1968] with FITC-labeled secondary antibodies to rabbit IgG (Sigma, United States). The corneas of adult Wistar rats and *P. waltl* newts were fixed in 4% formaldehyde solution in 0.01 M phosphate-buffered saline (PBS, pH 7.4) for 6 h, washed with PBS for 24 h at 4°C, dehydrated, and embedded in paraffin. Transverse histological sections (6–7 μm) were prepared for immunohistochemical analysis and treated with the anti-CRP antiserum (1 : 10) and FITC-labeled secondary antibodies (1 : 35). Control sections were treated with (1) secondary antibodies alone or (2) nonimmune rabbit serum. Immunofluorescence was analyzed and recorded using equipment manufactured by Olympus (Japan) and Leica (Germany).

In experiments on the effect of CRP on cell proliferation in the newt cornea *in vivo,* newts were anesthetized with MS 222, and the cornea was transversely incised with a sharp blade so that the wound length was equal to the diameter of the expanded pupil. During the following four weeks, the newts were immersed once daily in an CRP solution (10^{-11} mg protein/ml) for 10 min (newts of the control group remained in dechlorinated tap water). On days 6, 13, and 27 after surgery, two newts from each group were injected with a ^{3}H-thymidine solution (10.5 Ci/mmol; Isotope, Moscow) at a dose of 5 μCi/g body weight. On the next day, they were sacrificed to perform a radiographic analysis of proliferation and determine the index of labeled nuclei (ILN) in the cornea.

In experiments with *in vitro* cultures, newt corneas were isolated together with the limbus (series 1) or without it (series 2) and placed in dark glass vials with the culture medium for amphibians: 64% medium 199, 9% fetal calf serum, 27% deionized water, 1 M HEPES (150 μg/ml), and gentamicin sulfate (80 mg/ml). In test cultures, the medium was supplemented with CRP to a final concentration of 10^{-11} mg/ml (in control cultures, the equivalent volume of the medium was added). The vials were installed in an Assistant RM5 roller (Germany) adjusted to 35 rpm and incubated at 20–22°C for 7 days without replacing the medium. Experiments with rats were performed in the same way, but the cornea was cultured in medium 199 with gentamicin sulfate.

After seven days, the cultures were removed from the medium and fixed in the Bouin solution. Histological sections (6–8 μm) were stained with Caracci hematoxylin and eosin and studied under a microscope at a magnification of 15 × 40×. Using an ocular micrometer, the relative thickness and number of epithelial layers in each section were determined. The results were processed statistically using the Origin program.

Results and Discussion

Isolation of RPs from the bovine cornea and their purification followed the procedure developed for analyzing RPs displaying biological activity in ultra low doses [Yamskova et al., 1977; Yamskov et al., 1999]. The method of extraction excluded mechanical or enzymatic tissue disintegration, and the extract contained mainly secreted proteins and extracellular matrix proteins [Shashoua et al., 1984]. As shown previously, RPs in tissue extracts are in the form of intermolecular aggregations [Yamskova and Reznikova, 1991] and complexes with proteins that modulate their biological activity [Yamskova et al., 2004].

According to the results of SDS-electrophoresis in 12% PAAG, the apparent molecular weight of CRP is no more than 10 kDa (Figure 1). The corresponding low-molecular fraction was eluted from the gel after nondenaturing electrophoresis and tested for biological activity. In the cell adhesion test, it proved to be active at concentrations as low as 10^{-9}–10^{-12} mg protein/ml.

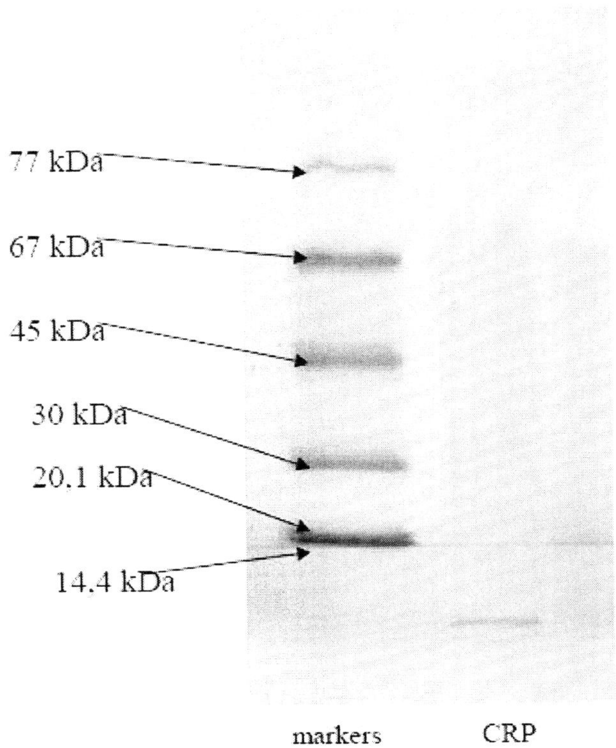

Figure 1. Electrophoregram of extract from the bovine cornea in 12% PAAG with sodium dodecylsulfate.

As inferred from circular dichroism spectra, the secondary structure of CRP is characterized by the prevalence of β-sheet regions (58%), whereas the content of α-helices is only about 14%. The remainder of the molecule (28%) has no definite structure (a random coil). These parameters correspond to those of other RPs identified in our studies. In addition, the presence of extensive β-sheet regions is indicative of the ability to form intermolecular aggregations [Ross and Poirier, 2004], which is also characteristic of these RPs. Indeed, according to the dynamic light scattering data, CRP in aqueous solutions forms nanoparticles with a hydrodynamic radius of 120±11 nm (Figure 2).

CRP was used as an antigen for raising specific rabbit antibodies, which, in turn, were used for its immunohistochemical localization in newt and rat corneas. In the cornea of *P. waltl* newt, CRP was detected mainly in the intercellular space of the endothelial layer and, in small amounts, in the epithelium [Margasyuk et al., 2005b] (Figure 3). The distribution of CRP in the rat cornea was similar, with fluorescence intensity being higher in the epithelium (Figure 4).

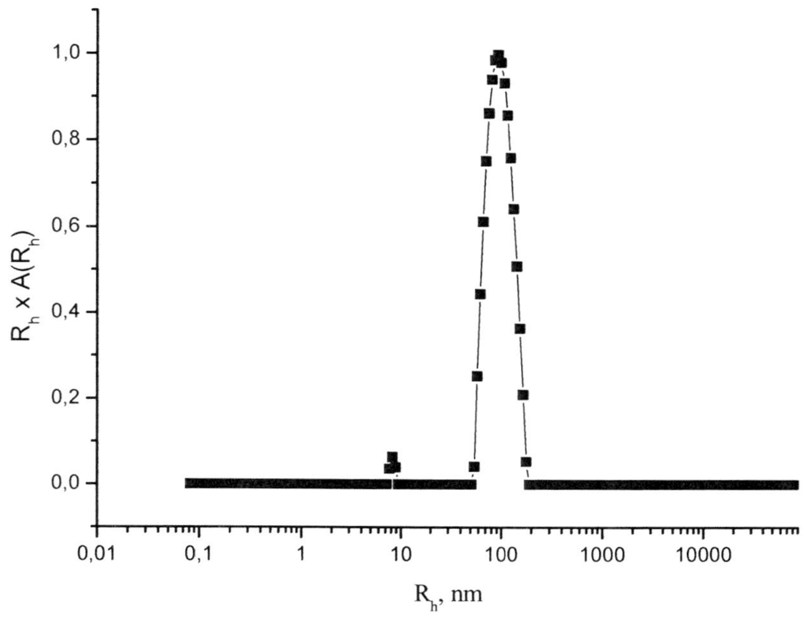

Figure 2. Distribution of CRP particles by hydrodynamic radius at 90°.

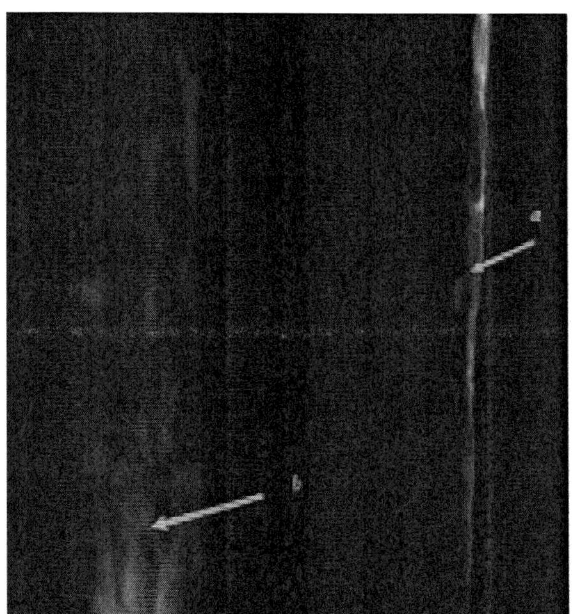

Figure 3. CRP localization in the normal newt cornea (arrow *a* – endothelium, *b* - epithelium).

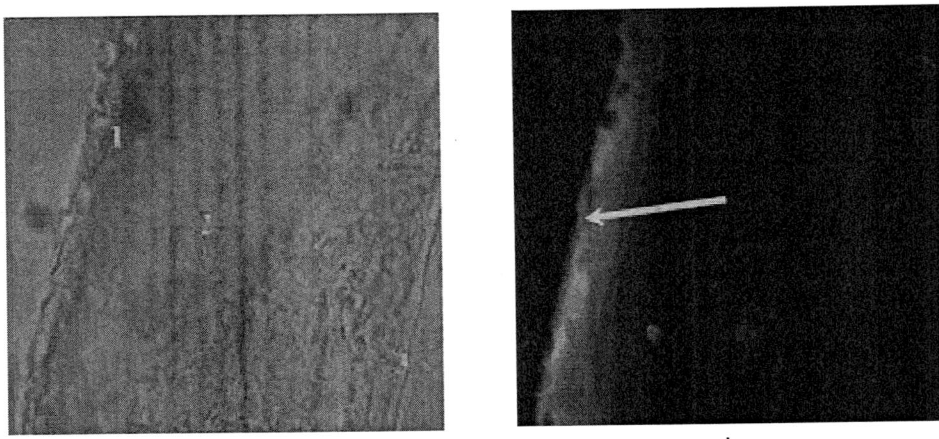

a b

Figure 4. CRP localization in the normal rat cornea. (a) Microphotograph of transverse histological section: (*1*) epithelium, (*2*) stroma, and (*3*) endothelium. (b) Immunohistochemical localization of CRP (arrow). Magnification: ocular 10×, objective lens 10×.

In rat corneas cultured for seven days, specific fluorescence in the intercellular spaces of endothelium and, especially, epithelium became markedly stronger than in preparations of the normal cornea (Figure 5). These results suggest that the CRP content in rat cornea cultures increases with time, which may be due either to intensification of CRP expression in cells of the corresponding corneal layers or to the release of CRP from a complex with other molecules in which it is unrecognizable to polyclonal antibodies. Anyway, it should be noted that the conditions of *in vitro* organ culture were the factor accounting for the above phenomenon.

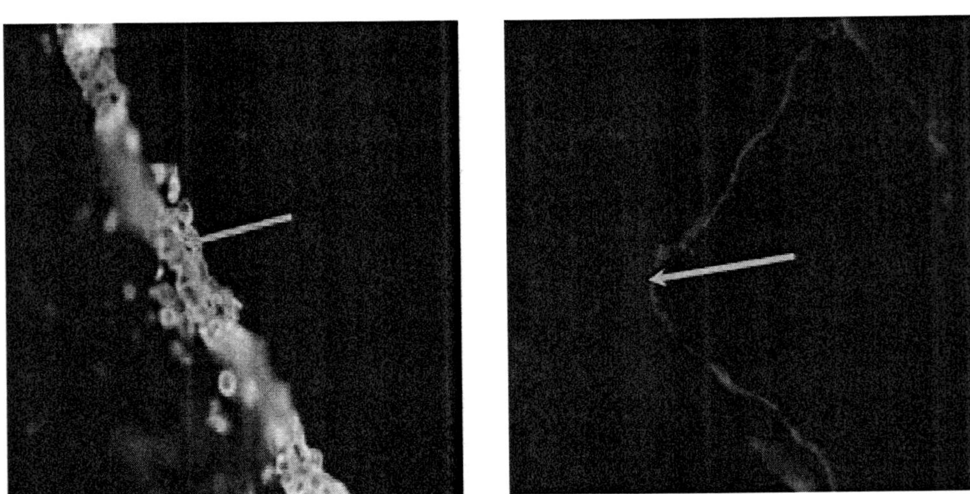

a b

Figure 5. CRP localization in (a) epithelium and (b) endothelium of the rat cornea in seven-day organ culture (arrows). Magnification: ocular 10×, objective lens 20×.

The effect of CRP on cell proliferation was studied on the model of corneal wound healing in newts. Daily treatment with CRP at ultra low doses proved to stimulate proliferation in the corneal epithelium (Figure 6). This effect reached a peak on day 7, which was reflected in a statistically significant increase of ILN relative to the control level ($p <$ 0.05). The stimulating effect of CRP was weaker but still significant ($p < 0.05$) on day 14, but it was not recorded on day 28 [Margasyuk et al., 2005a]. These results show that CRP stimulates cell proliferation in the corneal epithelium in the course of wound healing *in vivo.* Moreover, they apparently reflect the general dynamics of proliferation in the newt cornea under such conditions. The gradual decrease of ILN in all groups of newts during the experiment provides evidence for the reduction of proliferating cell pool in the cornea because of completion of its regeneration. It should also be noted that the significant increase of ILN ($p < 0.05$) relative to that in the normal cornea was observed not only in the wound zone, but also in sites that did not suffer direct damage, with differences in ILN between damaged and undamaged regions being statistically nonsignificant ($p <$ 0,05). Apparently, tissue repair in the injured cornea is a generalized process, with the "repair field" extending beyond the zone of direct injury. This assumption agrees with the concepts of morphogenetic fields [Gilbert et al., 1977, Beloussov, 2001] and related regeneration fields [Chiang et al., 1992; Robinson and Messerli, 2003].

At the nest stage of studies on the specific biological activity of CRP, we analyzed its protective effect on newt and rat corneas in seven-day roller cultures. The corneas for culturing were dissected either together with the limbus or without it.

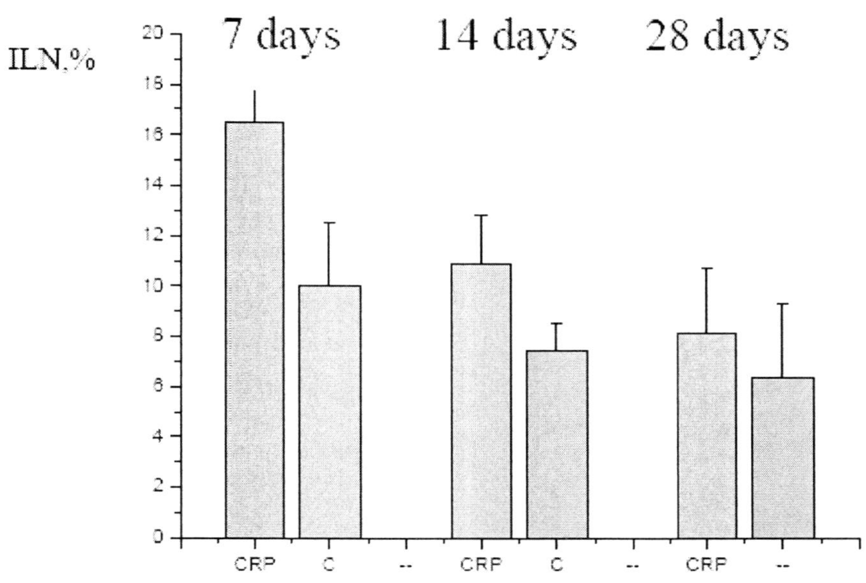

Figure 6. Changes in the index of labeled nuclei (ILN, %) in the newt corneal epithelium during wound healing *in vivo:* (CRP) corneas treated with CRP, (C) control corneas.

Cultures of the newt cornea with the limbus curled up and formed spherical structures consisting of three cell layers. The external and middle layers were natural (corneal epithelium and stroma, respectively), whereas the inner layer developed as a result of cell migration from the epithelium to the inner surface of the cornea and their subsequent proliferation, so that these cells substituted the endothelium. In cultures treated with CRP (10^{-11} mg/ml), the stratified structure of corneal epithelium was preserved significantly better than in control cultures without CRP (Fig. 7). In addition, these cultures differed in the thickness of each cell layer relative to the total thickness of the cornea taken as 100% (Table 1). Differences in these parameters were statistically significant at $p < 0.05$.

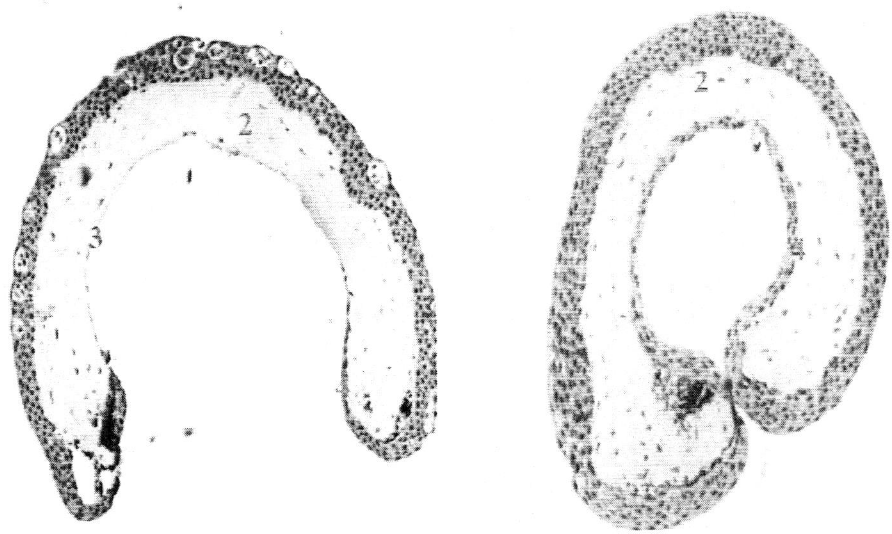

Figure 7. Seven-day organ cultures of the newt cornea with limbus: (a) control culture, (b) culture treated with CRP (*1*, outer (natural) epithelial layer; *2*, stroma; *3*, endothelium; *4*, inner epithelial layer). Magnification: ocular 4×, objective lens 10×.

Table 1. Parameters of the newt cornea after seven-day roller organ culturing

Parameter	Normal cornea	Cornea with limbus in culture		Cornea without limbus in culture	
		experiment	control	experiment	control
Relative thickness of outer epithelial layer, %	32.13 ± 1.08	36.19 ± 3.13	29.94 ± 4.91	35.89 ± 5.41	8.48 ± 2.46
Relative thickness of stroma, %	66.88 ± 2.14	58.98 ± 3.47	70.06 ± 4.48	64.11 ± 2.35	91.52 ± 2.83
Relative thickness of endothelium, %	0.99 ± 0.17	—	—	—	—
Relative thickness of inner epithelial layer, %	—	4.83 ± 1.71	—	—	—
Number of cell layers in epithelium	3.65 ± 0.58	5.15 ± 0.72	4.78 ± 1.71	3.46 ± 0.50	1.34 ± 0.48

A different picture was observed in cultures of the newt cornea without the limbus. The control cultures did not form spherical structures, epithelial cell adhesion to the stroma was disturbed, and the inner epithelial layer was absent. A large proportion of epithelial cells died, and the degree of epithelium degradation in many cultures was so high that the corresponding layer was virtually absent. In CRP-treated cultures, conversely, the epithelial layer consisted mainly of viable cells, although their adhesion to the stroma was also disturbed (Figure 8). As in the previous series, differences between these groups of cultures in the relative thickness of corneal layers and the number of cell layers in the epithelium were statistically significant at $p < 0.05$) (Table 1).

In experiments with the rat cornea, cultures with and without the limbus in the control and CRP-treated groups were spherical and consisted of two cell layers, epithelial and stromal. The inner, endothelial layer degraded and was absent in some regions. None of the cultures had an inner epithelial layer formed by cells migrating from the epithelium, as was the case with newt cornea + limbus cultures. Likewise, neither control nor CPR-treated cultures showed any disturbances of epithelial cell adhesion to the stroma. Thus, the state of the cornea in these two groups of cultures was largely similar. In statistical terms, the values of relative layer thickness in cornea+ limbus cultures (Figure 9) were similar in the control and CPR-treated groups, and only the difference in the number of cell layers in the epithelium was significant ($p < 0.05$). In corneas cultured without the limbus, conversely, differences both in the relative thickness of the epithelium and stroma and in the number of cell layers in the epithelium (Figure 10) between control and CRP-treated cultures were statistically significant. These data are presented in Table 2.

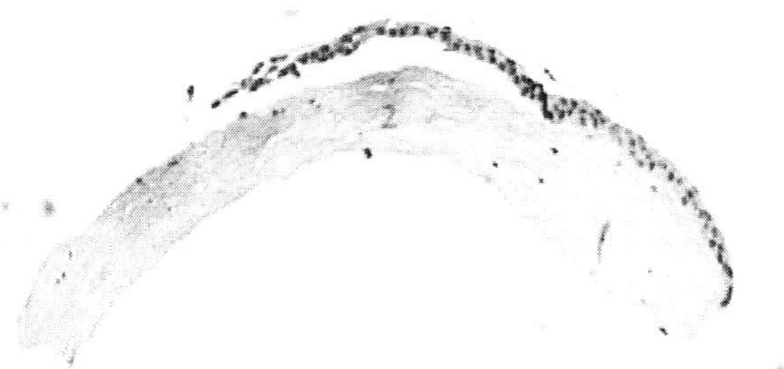

a

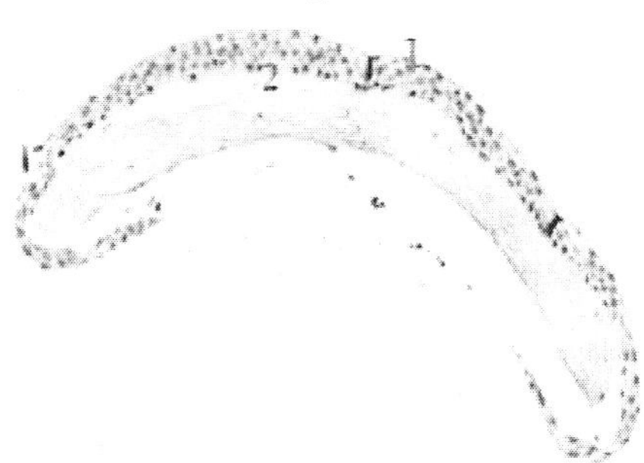

b

Figure 8. Seven-day organ cultures of the newt cornea without limbus: (a) control culture, (b) culture treated with CRP (*1*, epithelium; *2*, stroma). Magnification: ocular 4×, objective lens 10×.

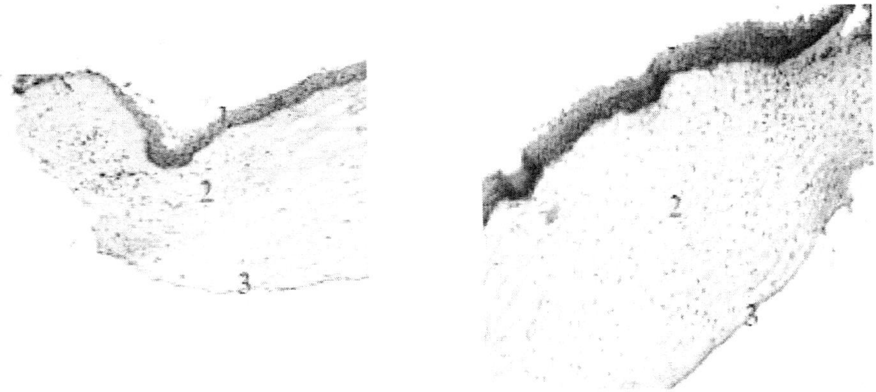

a b

Figure 9. Seven-day organ cultures of the rat cornea with limbus: (a) control culture, (b) culture treated with CRP (*1*, epithelium; *2*, stroma; *3*, endothelium). Magnification: ocular 10×, objective lens 10×.

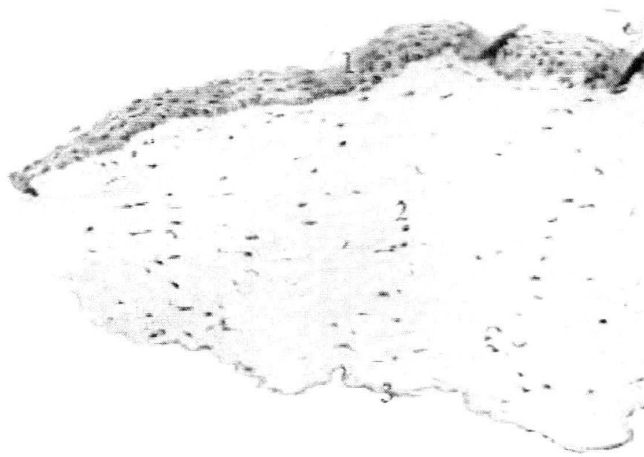

a

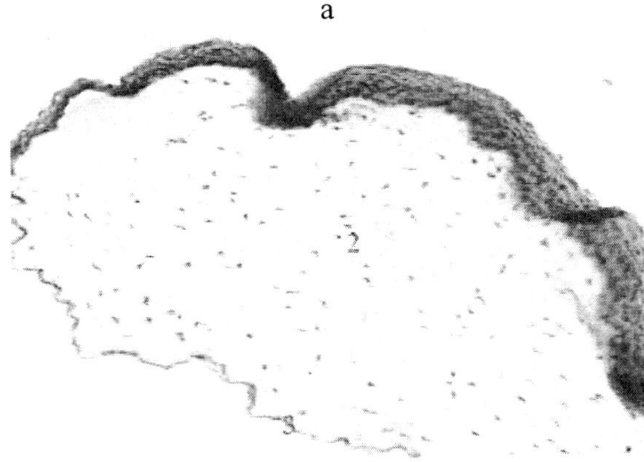

Figure 10. Seven-day organ cultures of the rat cornea without limbus: (a) control culture, (b) culture treated with CRP (*1*, epithelium; *2*, stroma; *3*, endothelium). Magnification: ocular 10×, objective lens 10×.

Table 2. Parameters of the rat cornea after seven-day roller organ culturing

Parameter	Normal cornea	Cornea with limbus in culture		Cornea without limbus in culture	
		experiment	control	experiment	control
Relative thickness of outer epithelial layer, %	20.03±3.90	11.91±1.63	11.69±2.41	12.67±0.91	15.93±3.01
Relative thickness of stroma, %	77.28±3.75	87.65±2.81	87.35±2.29	86.71±0.97	83.11±1.83
Relative thickness of endothelium, %	1.77±0.13	0.44±0.15	0.96±0.08	0.62±0.21	0.96±0.09
Relative thickness of inner epithelial layer, %	4.69±0.13	5.8±0.57	5.19±0.39	6.21±0.50	4.39±0.30
Number of cell layers in epithelium	20.03±3.90	11.91±1.63	11.69±2.41	12.67±0.91	15.93±3.01

It is noteworthy that the values of relative thickness of corneal layers in all cultures significantly differed from those in the normal cornea ($p < 0.05$). Although quantitative differences in this parameter were absent, it should be noted that the state of the stroma in cornea + limbus cultures treated with CRP differed from that in the control cultures: its lamellar structures was much closer to the structure of the normal cornea, with the arrangement of lamellae being more ordered.

Thus, the results obtained with organ cultures of the rat and newt cornea are largely similar. In both cases, the number of cell layers in the epithelium and its relative thickness increased under the effect of CRP. One of the factors responsible for changes in the relative thickness of corneal layers is degradation of the endothelium, because it is this layer that regulates water content in the stroma [Leuenberger, 1978, Bonanno, 2003]. Endothelial cell death or substitution by epithelial cells may provide for a higher water content and, therefore, greater thickness of the stroma, with a consequent decrease in the thickness of the epithelium. It should be noted that the relative thickness of the stroma increased in all cultures of the newt and rat cornea, compared to the normal tissue. However, it was always smaller in CRP-treated cultures than in control cultures. Apparently, CRP improved the viability of endothelial cells and thereby provided for a lower water content in the stroma. In addition, it stimulated proliferation and migration of epithelial cells, as the number of cell layers in the epithelium of CRP-treated cultures was greater than in the control.

Differences in the parameters of corneas cultured with and without the limbus deserve special attention. It may well be that CRP acts primarily upon epithelial cells of the limbus, which is one of the main sources of cells for cornea regeneration in amphibians and mammals [Davanger and Evensen, 1971, Cotsarelis et al., 1989], and both the increase in the number of cell layers in the epithelium and the formation of a new, internal epithelial cell layer in cultures of the newt cornea cultures are due to the stimulating effect of CRP on proliferation and migration of these cells. However, culturing of the rat cornea with the limbus did not entail any significant increase in the relative thickness of the cornea and the number of cell layers in the epithelium, compared to cultures without the limbus. Maybe, the progenitor cells of the rat limbus are less responsive to CRP or have a lower regeneration potential than limbus cells of the newt.

Thus, the results of our study show that the cornea contains a regulatory protein that exhibits biological activity at ultra low doses, and this activity lacks species specificity. This corneal regulatory protein (CRP) is located in the intercellular space of the corneal epithelium and endothelium. It stimulates proliferation and migration of corneal epithelial cells under conditions of *in vitro* organ culture and corneal wound healing *in vivo.*

REFERENCES

J. A. Bonanno, *Progress in Retinal and Eye Research*, Vol. 22, pp. 69–94. 2003.

E. I. Bueverova, E.V. Bragina, M.M. Reznikova, V.P. Yamskova, and N.G. Khrushchov, *Doklady Akademii Nauk SSSR*, Vol. 281, N 1, pp. 158–160, 1985.

L. V. Beloussov, Developmental Biology, Vol. 94, N 2, pp. 219-235, 2001.

M. Chiang, K.R. Robinson, J.W. Vanable, *Experimental Eye Research*, Vol. 54, N 6, pp. 999-1003, 1992.

G. Cotsarelis, G. Cheng, G. Dong, T.T. Sun, R.M. Lavker, *Cell*, Vol. 57, pp. 201– 209, 1989.

M. Davanger, A. Evenson, *Nature*, Vol. 229, pp. 560– 561, 1971.

N. V. Engelhardt, Immunokhimicheskii analiz (Immunochemical Analysis)//*Moscow: Meditsina*. 1968.

E. Engvall, P. Perlman, *Journal of Immunochemistry*, Vol. 8, pp. 871-874, 1971.

S. F. Gilbert, D.M. Opitz R.A. Raff, *Ontogenez,* Vol. 28, N 5, pp. 325–343, 1997.

"Immunoassays: a practical approach", Ed. By J. P. Gosling, New York, Oxford University Press. 2000. pp. 19 - 36.

B. D. Hames, "One-dimensional gel electrophoresis of proteins", in the book *"Gel Electrophoresis of Proteins. A Practical Approach"*, Ed. by B.D. Hames and D. Rickwood, New York, Oxford University Press, 1990, 106 pp.

M. S. Krasnov, E.N. Grigoryan, V.P. Yamskova, et al. *Ontogenez,* Vol. 31, N 5, pp. 368–373, 2000.

M. S. Krasnov, E.N. Grigoryan, V.P. Yamskova, Izvestia Rossiyskoy Akademii Nauk, *Seria Biologicheskaya*, N 1, pp. 22–36, 2003.

P. M. Leuenberger, *Adv. Ophtalmol.* 1978. vol. 35. pp. 94 – 166.

D. V. Margasyuk, E.N. Grigoryan, V.P. Yamskova, Izvestia Rossiyskoy Akademii Nauk, *Seria Biologicheskaya*, N 6, pp. 738–743, 2005a.

D. V. Margasyuk, M.S. Krasnov, V.P. Yamskova, E.N. Grigoryan, I.A. Yamskov, *Oftal'mologiya,* Vol. 2, N 3, pp. 81–87, 2005b.

S. W. Provencher, *Macromol. Chem*. Vol. 82 N 15, pp. 632, 1985.

K. R. Robinson, M.A. Messerli, *Bioassays,* Vol. 25, N 8, pp. 759-766, 2003.

C. A. Ross, M.A. Poirier, *Nat. Med.* 10 Suppl, pp. 10 - 17, 2004.

V. E. Shashoua, G.W. Hesse, B.W. Moore, *Journal of Neurochemistry*, Vol. 42, N 6, pp. 1536-1541, 1984.

P. K. Smith, R.I. Krohn, G.T. Hermanson, A.K. Mallia, F.H. Gartner, M.D. Provenzano, E.K. Fujimoto, N.M. Goeke, B.J. Olson, D.C. Klenk, *Analitical Biochemistry*, Vol. 150, N 1, pp. 76-85, 1985.

P. Stepanek, "Dynamic Light Scattering", in the book *"The method and some applications"*, Ed. by W. Brown, Oxford, Clarendron Press, 1993, 177 pp.

S. Yu Venyaminov, J.T. Yang, "Determination of protein secondary structure", in the book "*Circular Dichroismand the conformational analysis of Biomolecules*", Ed. by G.D.Fasman, New York, Plenum Press, pp. 69-107, 1996.

I. A. Yamskov, V.P. Yamskova, A.N. Danilenko, Z.S. Klemenkova, B.G. Antipov, F.R. Chernikov, M.M. Gusynina, E.Yu. Rybakova, *Rossiyskii Khimicheskii Zhurnal*, Vol. 43, N 5, pp. 34–39, 1999.

V. P. Yamskova, E.A. Modyanova, M.M. Reznikova, A.G. Malenkov, *Molecularnaya Biologia,* Vol. 11, N 5, pp. 1147–1154, 1977.

V. P. Yamskova, M.M. Reznikova, *Zhurnal Obshchey Biologii,* Vol. 52, N 2, pp. 181–191, 1991.

V. P. Yamskova, N.V. Nechaeva, N.B. Tumanova, Yu.G. Yurovitskii, T.E. Novikova, V.I. Fateeva, I.G. Gvazava, Izvestia Rossiyskoy Akademii Nauk, *Seria Biologicheskaya*, N 2, pp. 190–196, 1994.

V. P. Yamskova, I.A. Yamskov, *Rossiyskiy Khimicheskiy Zhurnal,* Vol. 43, N 2, pp. 74–79, 1999.

In: Biochemical Physics Frontal Research
Editors: Sergei D. Varfolomeev et al., pp. 61-70
ISBN 1-60021-425-8

Chapter 7

ANALYSIS OF REGULATORY PROTEINS FROM BOVINE BLOOD SERUM THAT DISPLAY BIOLOGICAL ACTIVITY AT ULTRA LOW DOSES: 1. ISOLATION, PURIFICATION AND PHYSICOCHEMICAL PROPERTIES

V. P. Yamskova, E. Yu. Rybakova, V. V. Vecherkin*, B. B. Berezin*, A. G. Filatova**, I. V. Blagodatskikh* and I. A. Yamskov*

Kol'tsov Institute of Developmental Biology, Russian Academy of Sciences
26 Vavilova str., Moscow, 117334 Russia
*Nesmeyanov Institute of Organoelement Compound,
Russian Academy of Sciences, 28 Vavilova str., Moscow, 117813 Russia
**N.M.Emanuel Institute of Biochemical Physics
Russian Academy of Sciences, 4 Kosygin str., Moscow, 11999, Russia

ABSTRACT

Three fractions containing regulatory proteins (RPs) biologically active at ultra low doses — weakly acidic, acidic, and basic — were isolated from bovine blood serum. The results of electrophoresis in PAAG showed that a low-molecular-weight protein ($R_f = 0.9 \pm 0.05$) was the main component of all these fractions. According to circular dichroism spectra, the secondary structure of RPs was characterized by the prevalence of β-sheet regions and unordered structures described in terms of random coil, with the content of α-helices being low. Tolerance of RPs to the impact of various physicochemical factors was noted in the course of the study. An analysis by the method of atomic power microscopy showed that RPs in solutions of different concentrations were in the form of spherical nanoparticles. The results of this study provide evidence that RPs from bovine serum are largely analogous to each other in their physicochemical properties.

Keywords: Blood serum, regulatory proteins, ultra low concentrations, physicochemical properties.

INTRODUCTION

Regulatory peptides occupy a special place among biologically active substances, as they provide for the maintenance of homeostasis at all levels of biological systems, from a single cell to the organism as a whole [1]. In addition to well-known signal molecules, such as peptide hormones, cytokines, and biologically active amines, this category includes low-molecular-weight proteins exhibiting biological activity at ultra low doses (RPs), which have been detected in plant and animal tissues [2–5]. These proteins are glycosylated, with the glycoside moiety consisting mainly of N-oligomannoside chains. Their secondary structure is characterized by the prevalence of β-sheet regions at a low content of α-helices, and a considerable part of the molecule is folded randomly (a stochastic globule). With respect to the general folding pattern, RPs are in the molten globule state [6]. Their molecules have a distinct tendency to combine into high-molecular-weight particles [7]. Ultra low concentrations of these proteins have an effect on basic biological processes, including cell adhesion, migration, proliferation, and differentiation, as well as stimulate regeneration and repair of pathologically altered tissues [8]. The biological activity of RPs is not species-specific but rather tissue-specific [9].

The blood serum RP is one of the best studied representatives of such proteins [7]. The RP from bovine blood serum has a molecular weight of about 12 kDa and pI in the range of 4.5 to 5.1 [6]. A 12-mer fragment of its molecule (molecular ion weight 1338.7 Da according to MALDI-TOF mass spectrometry) was sequenced [6].

It has been shown that the functional state of this RP in the blood serum changes during ontogeny: the RP is active in embryos, especially during the first two-thirds of the embryonic period, but gradually loses its activity by the end of embryonic development and remains inactive throughout postnatal life. Its inactivation is due to the formation of a complex with another serum protein [7, 10]. On the basis of available data, it has been proposed that such inactivation is a way to create reserves of this biologically active factor and control its activity in adult individuals [10].

We revealed two more RP fractions in the bovine blood serum and compared physicochemical properties of all serum RPs identified to date.

EXPERIMENTAL

Bovine serum—sterile, inactivated preparation for tissue cultures (15 l) — was from the Biological Preparations Division of the Moscow Meat-Packing Plant. All reagents were of chemical purity grade. Solutions were made in deionized water (15 MΩ) from a MilliQ column. Mice used in the study were F1 hybrid males weighing 20–22 g from the vivarium of the Kol'tsov Institute of Developmental Biology.

The procedure of RP isolation from the bovine serum involved two rounds of protein precipitation with ammonium sulfate followed by isoelectrofocusing (IEF) of the biologically active fraction [7]. Fractions obtained at each stage were checked for the presence of RP in a cell adhesion test developed for biological indication of proteins belonging to this group [7]. Protein concentration in these fractions was determined by the colorimetric method with bicinchoninic acid [11].

Three fractions were collected for further analysis: (1) acidic proteins with pI < 3.0, (2) proteins with pI 4.5–5.1, and (3) basic proteins with pI 8.5–9.5.

Reverse-phase HPLC was performed using a Yanaco chromatograph (Japan) with an Altex RPSC column (4.6 × 75 mm). Proteins were eluted with a 20–80% acetonitile concentration gradient containing 0.1% trifluoroacetic acid for 20 min, at a flow rate of 1 ml/min, with signal detection at 280 nm.

The protein composition of fractions was analyzed by SDS-electrophoresis in 12.5% PAAG in a Mini-Protean III cell (BioRad, Italy) as described [12]. Proteins used as molecular weight markers were as follows: phosphorylase b (94 kDa), bovine serum albumin (67 kDa), ovalbumin (45 kDa), carbonic anhydrase (30 kDa), soybean trypsin inhibitor (20 kDa), and chicken egg white lysozyme (14.4 kDa).

Nondenaturing electrophoresis in 12.5% PAAG was performed according to [14]. After electrophoretic separation of the active fraction obtained by IEF, the band corresponding to proteins with $R_f = 0.90 \pm 0.05$ was cut out, and proteins were eluted by incubating it in several small volumes of water. The eluates were pooled, dialyzed several times against distilled water, and lyophilized. The resultant preparation was analyzed by physicochemical methods.

Circular dichroism spectra were recorded in a Jasco 720 dichrograph (Japan) equipped with a computer data processing system, as described [3]. Their analysis was performed using the k2d program [18].

Experimental data were processed by methods of variation statistics. Student's *t*-test was used to estimate significance of differences.

RESULTS AND DISCUSSION

Proteins exhibiting biological activity at ultra low doses (RPs) have been detected in different tissues of vertebrates [2, 3, 6, 9]. According to experimental data, RPs in tissue extracts are represented by several fractions of acidic and basic proteins [4, 9]. In addition, weakly acidic RP with pI in the range of 4.5 to 5.1 was previously detected in blood serum [7, 10]. Taking into account these data, the bovine blood serum was fractionated using the procedure described above, and fractions of acidic and basic proteins collected after IEF were biologically tested for the presence of RPs. In addition, similar experiments were performed with the weakly acidic fraction (pI 4.5–5.1) (Figure 1).

The results of tests showed that acidic and basic protein fractions at ultra low doses had a membranotropic effect, as in the case of blood serum RP analyzed in our previous studies [7, 10] (Figure 2).

Thus, three fractions containing regulatory proteins (RPs) biologically active at ultra low doses - weakly acidic (WRP), acidic (ARP), and basic (BRP) have been investigated.

Figure 3 shows the protein pattern of three RP-containing blood serum fractions after electrophoresis in 12.5% PAAG under denaturing conditions. It is seen that the main component of each fraction is a low-molecular-weight protein (< 10 kDa). We made an attempt to determine the molecular weight of test proteins more accurately by separating them in 18% PAAG. However, this experiment failed, as no Coomassie-stained bands were found in the gel.

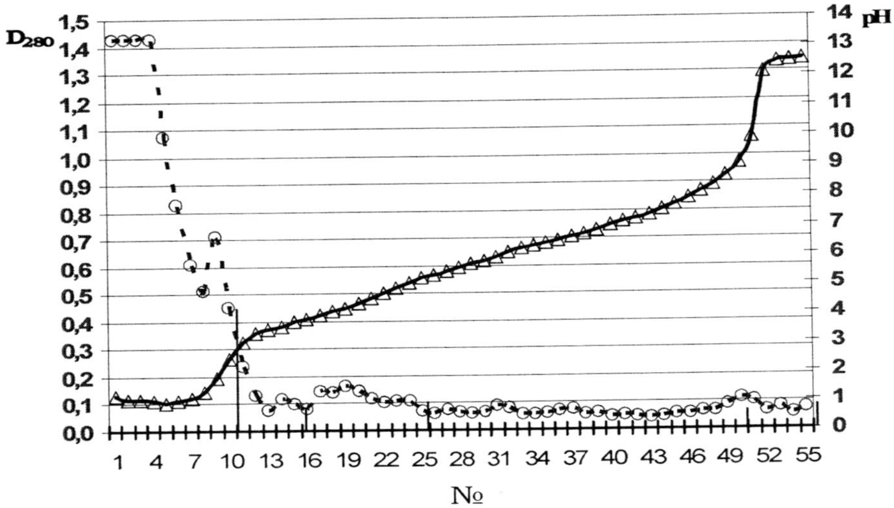

Figure 1. Pattern of fractions (1–3) after isoelectric focusing of proteins from bovine blood serum. Ordinates 1 and 2 shows pH and D_{280}, respectively; abscissa shows number of fractions.

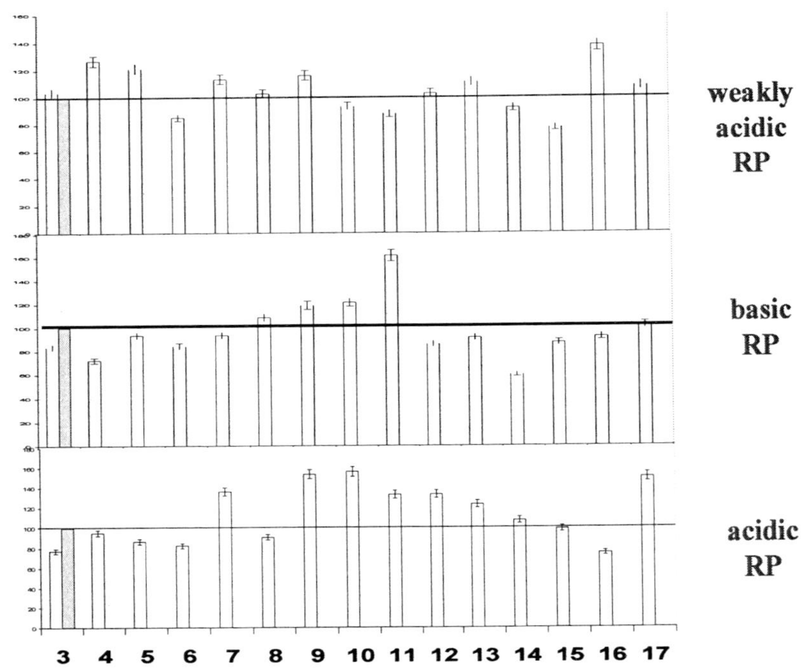

Figure 2. Activity of RP-containing blood serum fractions (black bars) purefied by IEF versus control (white bars) in biological tests Ordinate shows the magnitude of membranotropic effect [7]; abscissa shows a exponent of power (x) of dilution factors of RP preparation (initial concentration approximately 0.1 mg/ml).

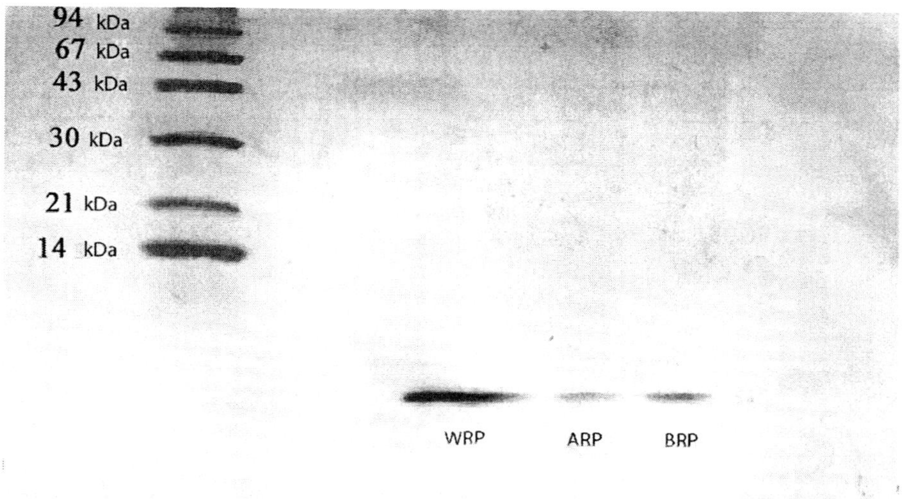

Figure 3. SDS-electrophoresis of (WRP) weakly acidic, (ARP) acidic, and (BRP) basic RP fractions from bovine blood serum.

Thereafter, we electrophoretically separated these fractions of blood serum in 12.5% PAAG under nondenaturing conditions and collected the corresponding low-molecular-weight proteins. In biological tests, all three preparations at ultra low doses displayed membranotropic activity.

Thus, RPs exhibiting biological activity at very low concentrations were represented in bovine blood serum by three low-molecular-weight protein fractions differing in electric charge.

These three RPs identified in the bovine blood serum were subsequently analyzed by the methods of reverse-phase HPLC and circular dichroism. It was found that the times of their retention on the chromatographic column were similar: 3.3 min for the acidic RP, 3.2 min for the weakly acidic RP, and 3.6 min for the basic RP. This is evidence that the degree of hydrophobicity of these RPs in all cases is virtually the same (Figure 4).

After HPLC, the acidic RP was analyzed by the method of Electrospray mass spectrometry. As a result, a peptide with a molecular ion weight of 1222 Da was identified. It is noteworthy that this value is close to the molecular ion weight of the blood serum RP revealed in the previous study [6]. Therefore, it is most probable that the RPs of interest are low-molecular-weight proteins.

An analysis of the secondary structure of blood serum RPs showed that, in all cases, it was characterized by the prevalence of β-sheet regions (mainly antiparallel structures and folds), with a considerable part of the molecule is folded randomly (Table 1). The circular dichroism spectra of all RPs changed in a similar way under the influence of physicochemical factors, such as heating or addition of EDTA, returning to the initial state when this influence ceased. This was additional evidence for close similarity of these RPs with respect to the secondary structure of their molecules.

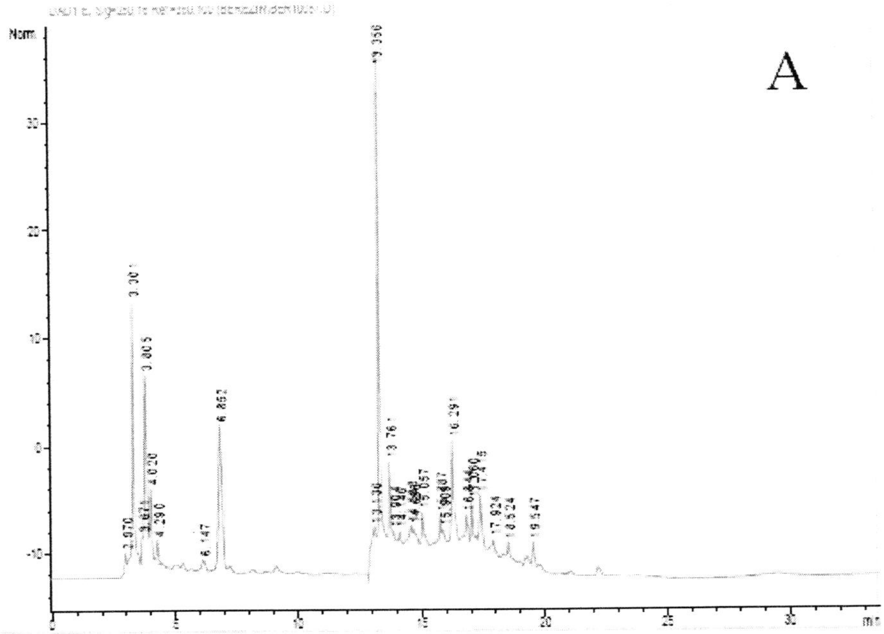

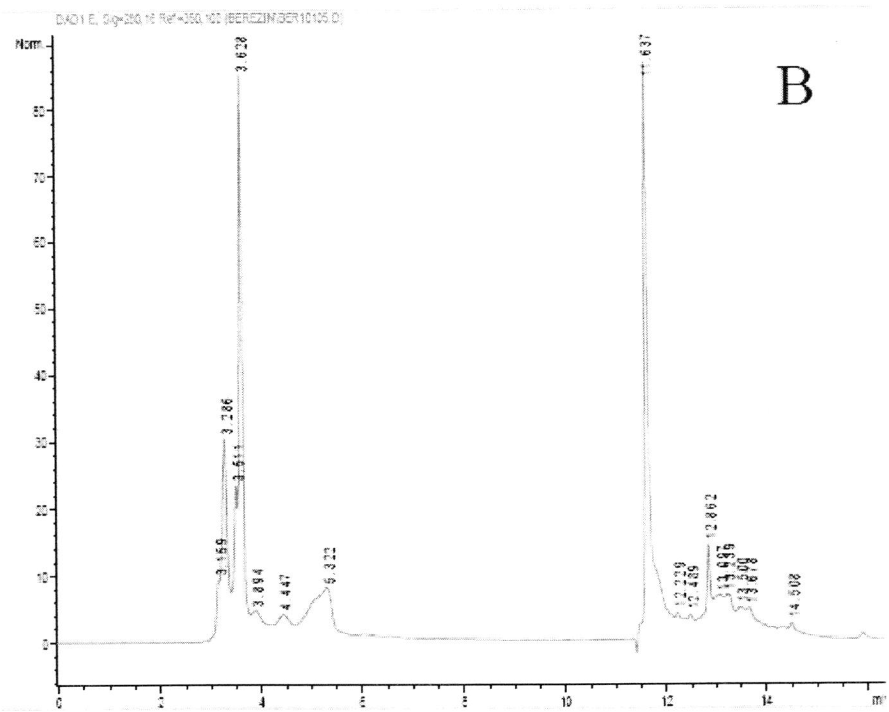

Figure 4 continued

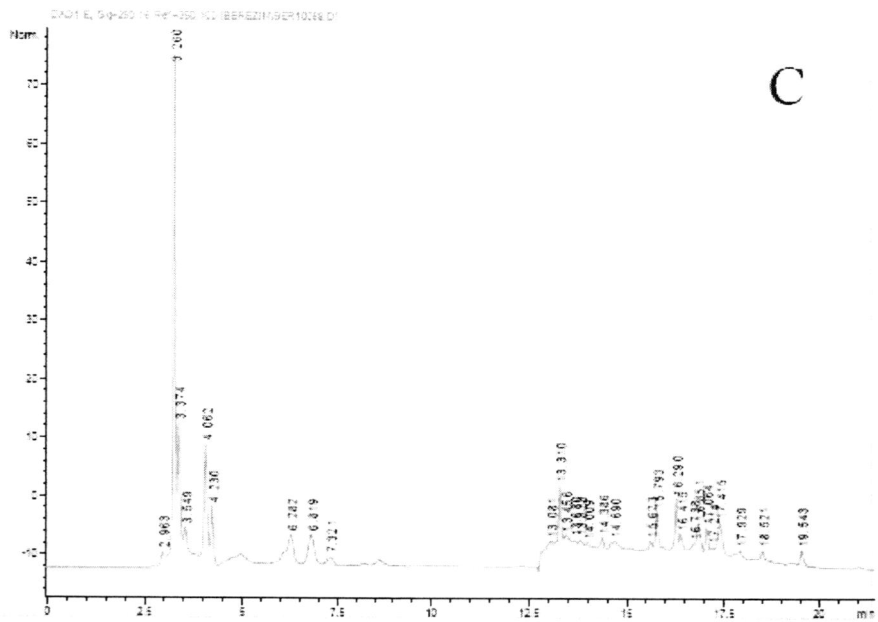

Figure 4. HPLC of (a) acidic, (b) weakly acidic, and (c) basic RP fractions from bovine blood serum.

Table 1. Parameters of the secondary structure of RPs from bovine blood serum

Elements of secondary structure	Content in RP molecules, %		
	weakly acidic RP	acidic RP	basic RP
α–Helices	10 ± 0.5	8 ± 0.5	9 ± 0.5
Antiparallel β–sheets	38 ± 0.5	40 ± 0.5	40 ± 0.5
Parallel β–sheets	2 ± 0.5	2 ± 0.5	2 ± 0.5
β–Folds	18 ± 0.5	17 ± 0.5	18 ± 0.5
Random coil	32 ± 0.5	33 ± 0.5	31 ± 0.5

Thus, the blood serum RPs proved to be exceptionally tolerant to the impact of physicochemical factors: their biological activity remained unchanged after exposure to heating or the action of EDTA. It should be noted, however, that the effects of these factors differed: unlike heating, EDTA as a helating agent caused considerable changes in the secondary structure of RPs, making it less ordered (Figures 5, 6). This can be explained by the fact that RPs are Ca^{+2}-binding proteins [14], and the conformation of such molecules is likely to change in the absence of calcium ions.

As follows from published data, protein molecules in which β–sheet regions prevail have a strong tendency to form intermolecular aggregations [15]. Indeed, dynamic light scattering (Figure 7) and atomic power microscopy revealed such aggregations of RP molecules in the form of spherical particles 150–200 nm in diameter (Figure 8). These particles proved to have a lipid component, which was extracted with organic solvents and assayed by MALDI-TOF mass spectrometry. As a result, decanoic, palmitic, linoleic, deoxycholic, and cholic acids were detected. The occurrence of serum RPs in the form of such particles may be one of the factors accounting for the high tolerance of these proteins to the action of physicochemical factors and enzymes.

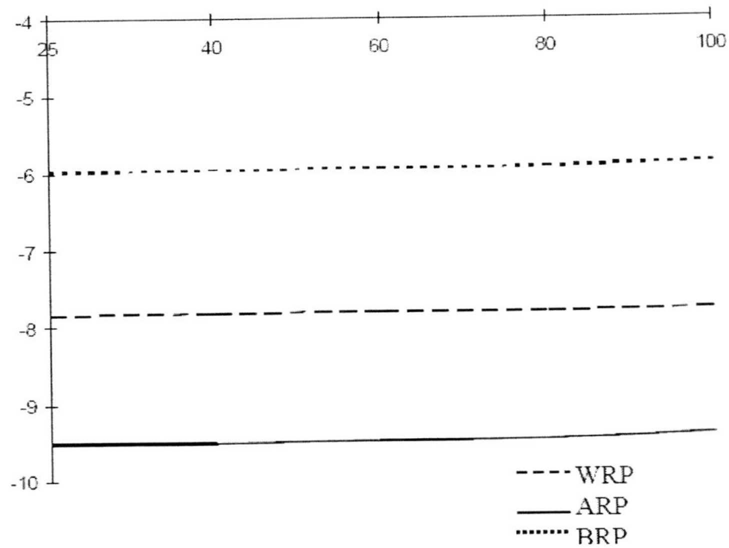

Figure 5. Circular dichroism spectra of blood serum RPs exposed to heating at 100°C for 30 min. Ordinate shows ellipticity at 222 nm; abscissa shows temperature of solution.

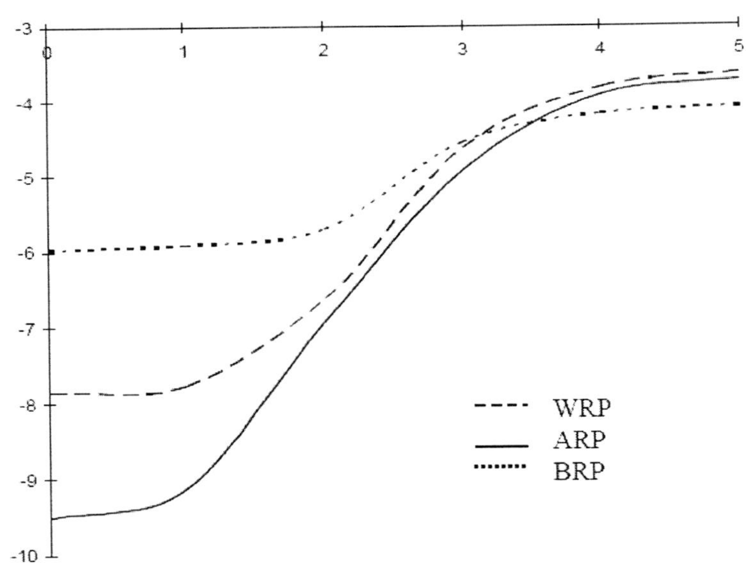

Figure 6. Circular dichroism spectra of blood serum RPs after adding EDTA. Ordinate shows ellipticity at 222 nm; abscissa shows EDTA concentration: (*1*) 0.2×10^{-2} M, (*2*) 0.5 (10^{-2} M, (*3*) 0.01 M, (*4*) 0.02 M, and (*5*) 0.05 M.

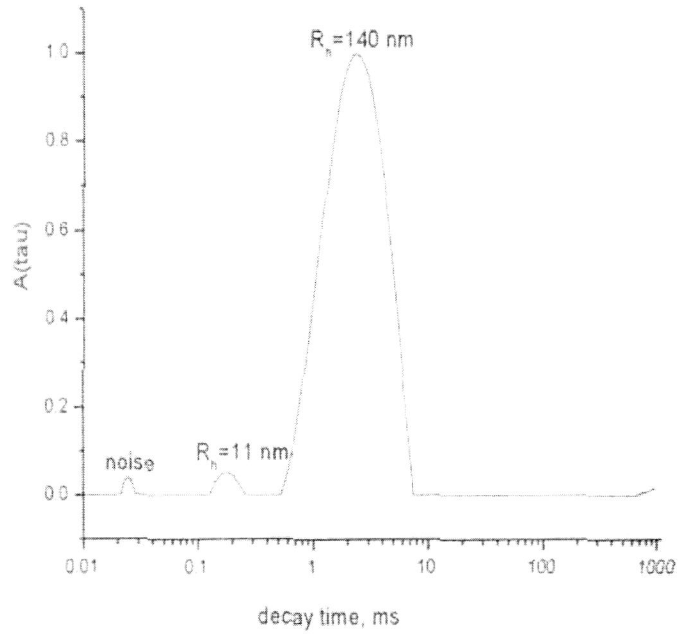

Figure 7. Dynamic light scattering by a solution of weakly acidic serum RP (helium–neon laser, λ = 633 nm; protein concentration 2×10^{-2} μg/ml). Ordinate shows scattering index; abscissa shows delay time, ms.

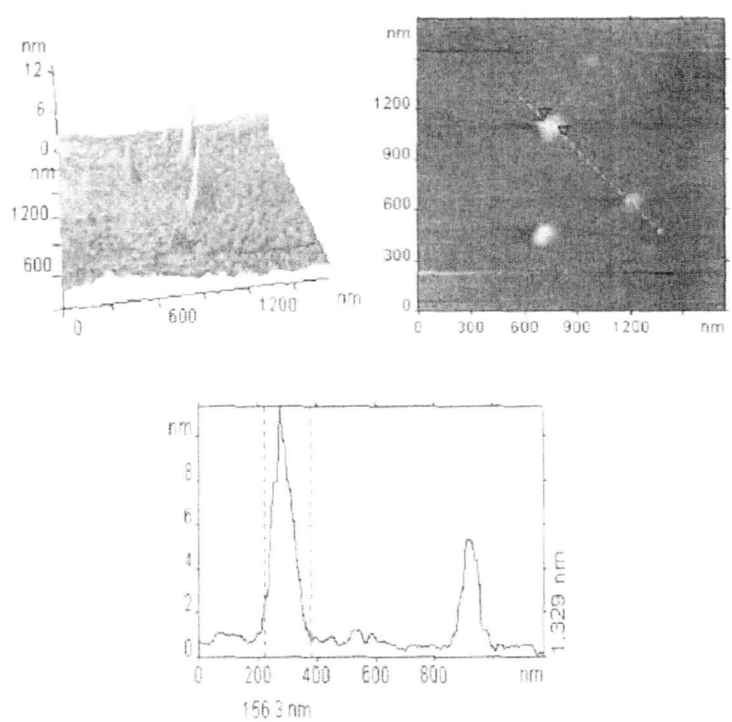

Figure 8. Atomic power microscopy of a solution of weakly acidic serum RP.

REFERENCES

[1] V. Kh. Khavinson, I.M. Kvetnoy, V.V. Yuzhakov, V.V. Popuchiev, S.S. Konovalov, *"Peptidergic regulation of homeostasis"*, St. Petersburg, "Nauka", 2003, 194 pp.

[2] V.P. Yamskova, E.A. Modyanova, V.I. Leventhal, T.P. Lankovskaya, O.K. Bocharova, A.G. Malenkov, *Biofizika,* Vol. 22, pp. 168 – 174, 1977a.

[3] V.P.Yamskova, E.A Modyanova, M.M.Reznikova, A.G.Malenkov, Molekulyarnaya. Biolologiya, Vol. 11, N 5,pp. 1147 – 1154, 1977b

[4] I.A.Yamskov, V.P.Yamskova, A.N.Danilenko, Z.S Klemenkova, B.G. Antipov, F.R Chernikov, M.M.Gusynina, E.Yu. Rybakova, *Rossiyskiy Khimicheskiy Zhurnal* (Zhurnal Rossiyskogo Khimicheskogo Obshchestva im. D.I. Mendeleeva), Vol. 43, N 5, pp. 34–39, 1999.

[5] M.S. Krasnov, D.V. Margasyuk, I.A. Yamskov, V.P. Yamskova, *Radiat. Biologiya Radioekologiya,* N 3, pp. 269 – 272, 2003a.

[6] I.A. Yamskov, A.A. Vinogradov, A.N.Danilenko, L.A. Maslova, V.P. Yamskova, *Prikladnaya Biokhimiya I Mikrobiologiya*, Vol. 37, N 1, pp. 36 -42, 2001.

[7] V.P.Yamskova, M.M. Reznikova, *Zhurnal Obshchey Biologii,* Vol. 52, N 2, pp. 181–191, 1991.

[8] R.A. Gundorova, I.P Khoroshilova-Maslova, E.V. Chentsova, L.V.Ilatovskaya, V.P. Yamskova, I.Yu Romanova, *Vestnik Oftal'mologii*, Vol. 113, N 2, pp. 12–15, 1997.

[9] M.S. Krasnov, E.N. Grigoryan, V.P.Yamskova, Izvestiya Rossiyskoy Akademii Nauk, *Seriya Biologicheskaya*, N 1, pp. 22 – 36, 2003b.

[10] V.P. Yamskova**,** E.Yu. Rybakova, A.A. Vinogradov, V.V. Vecherkin, I.A. Yamskov, *Prikladnaya Biokhimiya I Mikrobiologiya*, Vol. 40, N 4, pp. 407-413, 2004.

[11] P.Smith, *Anal. Biochem*., Vol.150, № 1, pp.76 – 85, 1985.

[12] U.K. Laemmli, *Nature,* Vol. 227, pp. 680–685, 1970.

[13] B.D.Hames"One-dimensional gel electrophoresis of proteins" in the book *"Gel Electrophoresis of Proteins. A Practical Approach*", Eds by B.D. Hames and D. Rickwood, New York, Oxford University Press, 1990, 106 pp.

[14] M.S. Krasnov, E.N. Grigoryan, V.P. Yamskova, D.V. Bogyslavskiy, I.A. Yamskov, *Radiat. Biolologiya Radioekologiya,* N 3, pp. 265 -268, 2003.

[15] C.A. Ross, M.A. Poirier, *Nat. Med*., 10 Suppl., pp. 10 – 17, 2004.

[16] D.V. Margasyuk, M.S. Krasnov, V.P. Yamskova, E.N. Grigoryan, I.A. Yamskov, *Oftal'mologiya*, Vol. 2, N 3, pp. 81 – 87, 2005.

[17] B. Chu, *"Laser Light Scattering : basic principles and practice"*, Boston, Academic Press, 1991.

[18] S.Yu. Venyaminov, J.T. Yang. "Determination of protein secondary structure", pp.69-107, in the book "*Circular Dichroismand the conformational analysis of Biomolecules*", Ed. By G.D.Fasman, New York, Plenum Press, 1996.

[19] S. W. Provencher, *Macromolecula Chemistry*, Vol. 82, N 15, pp.632, 1985.

In: Biochemical Physics: Frontal Research
Editors: Sergei D. Varfolomeev et al. pp. 71-78
ISBN 1-60021-425-8

Chapter 8

ANALYSIS OF REGULATORY PROTEINS FROM BOVINE BLOOD SERUM THAT DISPLAY BIOLOGICAL ACTIVITY AT ULTRA LOW DOSES: 2. TISSUE LOCALIZATION AND ROLE IN WOUND HEALING

V. P. Yamskova, M. S. Krasnov, E. Yu. Rybakova, V. V. Vecherkin*, A. V. Borisenko and I. A. Yamskov*

N.K. Kol'tsov Institute of Developmental Biology,
Russian Academy of Sciences, 26 Vavilova str., Moscow 119991, Russia;
*N.A. Nesmeyanov Institute of Organoelement Compounds,
Russian Academy of Sciences, 28 Vavilova str., Moscow 117813, Russia

ABSTRACT

Regulatory proteins (RPs) detected in fractions of bovine blood serum-acidic (pI < 3.0), weakly acidic (pI 4.5–5.1), and basic (pH 8.5–9.5) - were analyzed for specific biological activity *in vivo* and *in vitro*. In addition, immunohistochemical localization of the weakly acidic RP in tissues was performed. This RP was detected in the intercellular space of the newt liver, which was indicative of its synthesis in this organ. All blood serum RPs at ultra low doses exhibited a protective effect on *in vitro* cultures of the newt skin. The weakly acidic RP (pI 4.5–5.1) proved to stimulate skin repair after injury, facilitating wound healing in mice.

INTRODUCTION

According to the results of our research on regulatory proteins (RPs) from bovine blood serum, the fraction containing a weakly acidic RP (pI 4.5–5.1) stimulates proliferation of fibroblasts *in vivo* [Bueverova et al., 1985] and has a favorable effect on corneal wound healing in rabbits [Gundorova et al., 1997]. The purpose of this study was to analyze the effects of three blood serum RPs on the state of newt skin explants cultured *in vitro* and on the healing of skin wounds in mice. Such an approach is appropriate in view of the fact that the biological effect of these RPs is not species-specific [Yamskova, Reznikova, 1991].

Material and Methods

Bovine serum - sterile, inactivated preparation for tissue cultures (15 l) - was from the Biological Preparations Division of the Moscow Meat-Packing Plant. Mice (F1 hybrid males weighing 20–22 g) and *Pleurodeles waltl* newts were from the vivarium (aquarium) of the Kol'tsov Institute of Developmental Biology. All reagents were of chemical purity grade. Solutions were made in deionized water (15 MΩ) from a MilliQ column.

RPs were isolated from the bovine blood serum according to the procedure used in our previous studies [Yamskova et al., 1977a, b]. Fractions obtained at each stage were checked for the presence of RP in a cell adhesion test developed for biological indication of proteins belonging to this group [Yamskova Reznikova, 1991]. Three fractions were collected: (1) acidic proteins with pI < 3.0, (2) weakly acidic proteins with pI 4.5–5.1, and (3) basic proteins with pI 8.5–9.5. They were additionally purified by nondenaturing electrophoresis in 12.5% PAAG [Hames, 1990].

Protein concentration in the fractions was determined by the colorimetric method with bicinchoninic acid [Smith et al., 1985].

Antisera for immunochemical RP identification were obtained by conventional methods [Mikhailov, Simirsky, 1991; Gosling, 2000]. Chinchilla rabbits weighing approximately 1 kg were immunized with the weakly acidic RP preparation (pI 4.5–5.1) isolated from the bovine blood serum as described [Yamskov et al., 2001]. The antigen solution was mixed with the equal volume of complete Freund's adjuvant, and each rabbit received 1 ml of the mixture injected under the skin in the shoulder region and into the femoral muscles at approximately equal doses. Thereafter, injections of the antigen mixed with incomplete Freund's adjuvant (1:1) were made in the same way at seven-day intervals. Blood samples taken from the marginal ear vein beginning from day 14 after the first injection, at the same intervals, were tested for the presence of specific antibodies by means of ELISA [Engvall, Perlman, 1971].

The localization and expression of the corresponding RP in the newt liver were studied by an indirect immunofluorescent method [Engelhardt, 1968], using the primary antiserum to the protein of interest diluted 1:10 and FITC-labeled secondary antibodies to rabbit IgG (whole molecule) (Sigma, United States) diluted 1:35. Control sections were treated with (1) secondary antibodies alone or (2) nonimmune rabbit serum. Immunofluorescence was analyzed and recorded using equipment manufactured by Olympus (Japan) and Leica (Germany).

Experiments with *in vitro* skin cultures were performed with mature *Pleurodeles waltl* newts (males and females) in autumn. The animals were anesthetized with 0.1% MS-222 solution and decapitated. The skin from the dorsal region was removed and cut into 5×5-mm squares, which were placed in dark glass vials with the culture medium (70 ml of medium 199, 10 ml of fetal calf serum, 30 ml of sterile deionized water, 30 μl 1M HEPES buffer and 200 μl of 4% gentamycin sulfate). The medium was sterilized by ultrafiltration through CA membrane filters (Nalgene, United States) with a pore size of 0.22 μm. The RP preparations were added to a final protein concentration of 10^{-10}–10^{-14} mg/ml of the medium. The control cultures were supplemented with the equivalent volume of sterile deionized water. The vials were closed with sterile stoppers fixed in place with Parafilm M film, installed in an RM 5 roller (Assistant, Germany) adjusted to 35 rpm, and incubated at 20–22°C. The medium was replaced on day 7, with the volume and concentration of RPs remaining the same. The cultures were fixed on days 7 and 14 with the Bouin solution. Histological section (3–4 μm) were stained with hematoxylin–eosin and analyzed under an Olympus microscope VANOX, AHBT3 (Japan).

Experiments on wound healing were performed with three groups of mice, five animals each, operated under ether anesthesia to remove a fragment of dorsal skin approximately 1 cm in diameter. In the first control group, the wound remained untreated. In the second control group, the wound was irrigated once a day with 100 μl of a saline. In the experimental group, the wound was treated once a day with 100 μl of a solution containing weakly acidic RP at a concentration of 1×10^{-10} mg/ml.

On day 11 after surgery, the mice were sacrificed, and tissue samples were fixed in the Bouin solution. Transverse sections through the wound region (7 μm) were stained with hematoxylin–eosin and viewed under an Olympus VANOX, AHBT3 microscope.

Experimental data were processed by methods of variation statistics. Significance of differences was estimated by Student's *t*-test.

Results And Discussion

For immunohistochemical localization of weakly acidic serum RP (pI 4.5–5.1), we used preparations of the newt liver, because the liver produces most of serum proteins [Engelhardt, 1984] and, in amphibians, is an organ of hematopoiesis in the postnatal period [Sichel et al., 2002]. In experiments with a specific antiserum to this protein, it was detected in the intercellular space between hepatocytes and on the surface of hematopoietic cells (Figure 1). This is evidence that the weakly acidic serum RP is indeed synthesized in the newt liver, and its occurrence in considerable amounts on the surface of hematopoietic cells is indicative of its probable involvement in hematopoiesis.

An analysis of the specific biological activity of serum RPs was a special task in view of the data obtained in our previous studies:

(1) The biological effect of these RPs lacks species specificity but is tissue-specific [Krasnov et al., 2003].
(2) The biological effect of RPs manifests itself only on condition that the spatial organization of cell microenvironment is preserved and does not differ significantly from that in normal tissue. Disturbances of contact interactions and cell adhesion result in the loss of this effect, even if the cells retain their location and arrangement in the tissue [Yamskova et al., 1990; 1994].
(3) The weakly acidic serum RP was shown to stimulate proliferation of mammalian fibroblasts *in vitro* [Bueverova et al., 1985] and markedly facilitate corneal wound healing in rabbits [Gundorova et al., 1997].

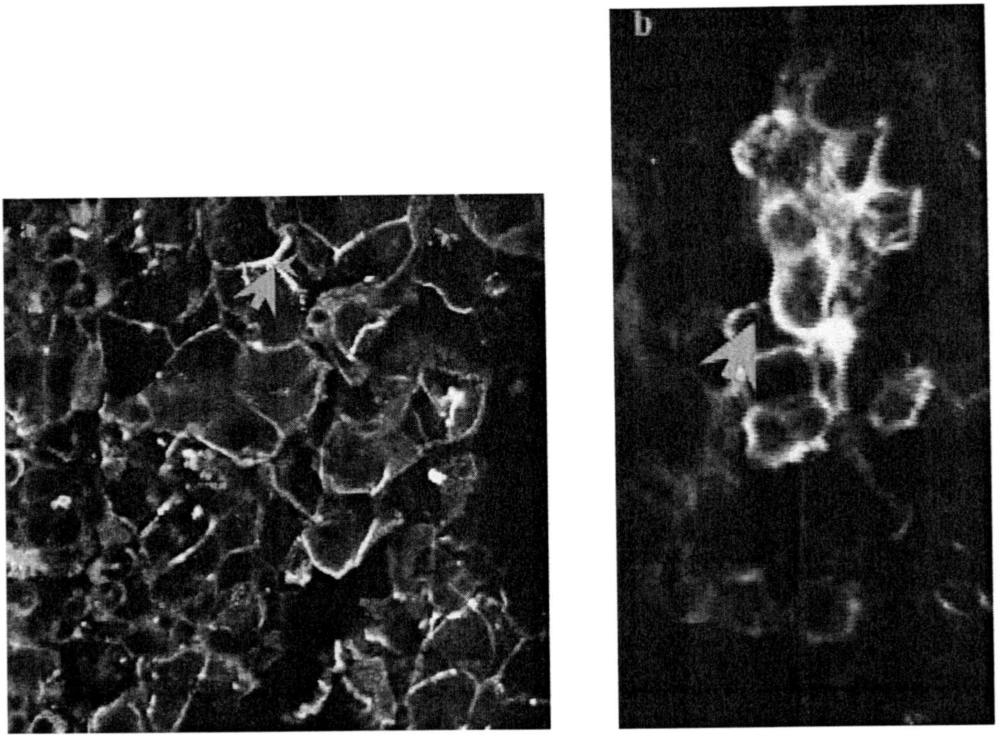

Figure 1. Immunohistochemical localization of RP (pI 4,5-5,1) (a) in the intercellular space between hepatocytes (arrow) and (b) on the surface of hematopoietic cells (arrow). Magnification: ocular 10×, objective lens 20×.

Among experimental *in vitro* models, roller organ cultures of the newt skin appeared to be promising for this purpose:

- Newt tissues have a greater regeneration potential than tissues of higher vertebrates [Grigoryan et al., 2005].
- In organ cultures, the spatial organization of cell microenvironment (intercellular space) in tissues is preserved.
- The method of roller cultures excludes the effect of tissue–substrate interactions.

Figure 2a shows a preparation of the normal newt skin with the stratified epidermis and the underlying three-layered dermis. The dermis contains numerous glands that secrete the mucous coating of the skin [Vorontzova et al., 1952].

On day 7, destructive processes were observed in control cultures. In particular, epidermal cells deteriorated, and skin glands were filled with mucous secretion (Figure 2b). In the cultures treated with the acidic RP, the surface of skin fragments was covered with almost normal stratified epithelium and had a thick mucous coating (Figure 2c). Numerous pigmented cells concentrated at the epidermis–dermis boundary. It is known that the number of such cells increases in response to exposure under unfavorable conditions [Riley, 1997]. Fibroblasts were detectable in the dermis. The cultures treated with the other two RP fractions were in the same state.

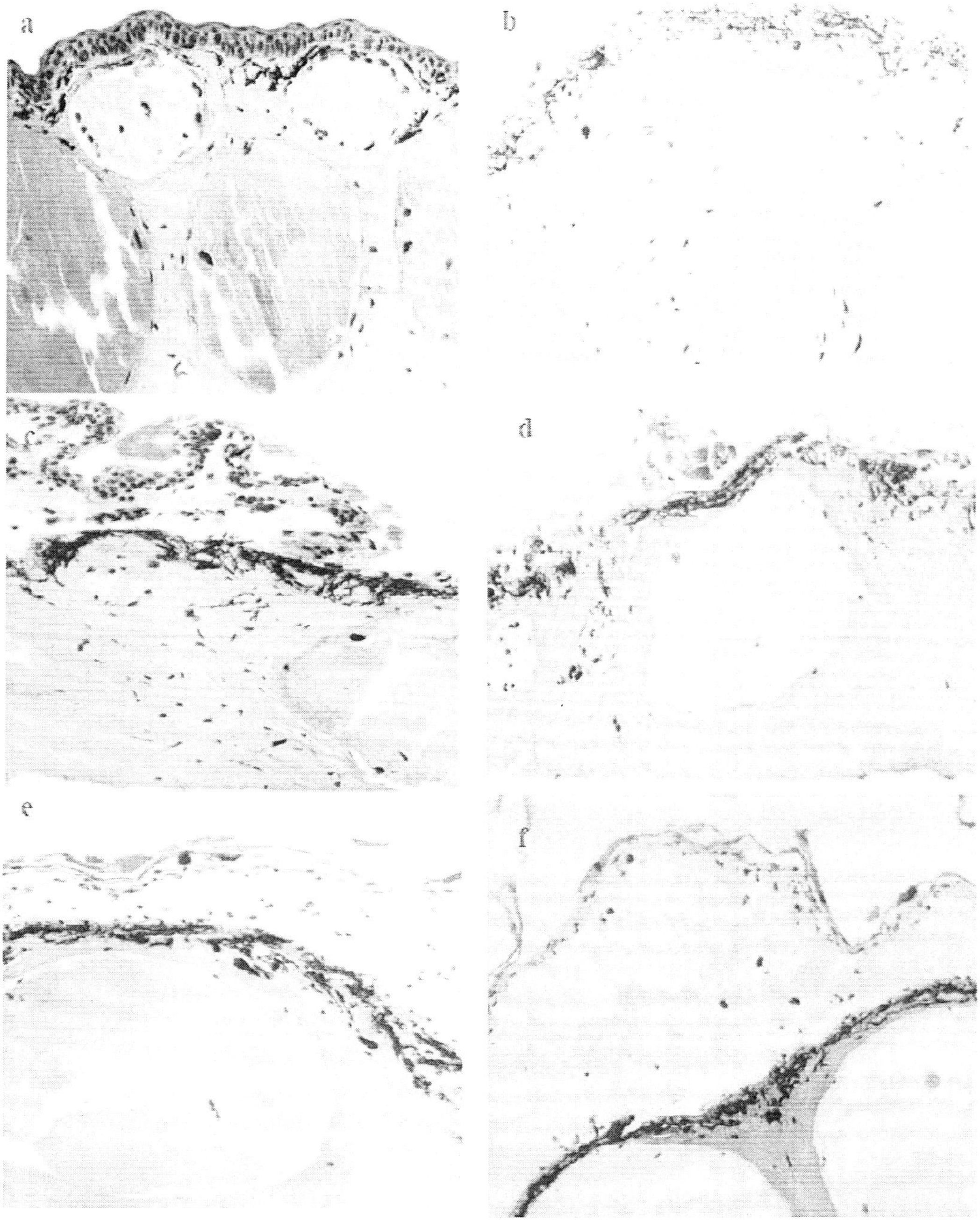

Figure 2. Transverse section of the newt skin (a) the normal newt skin; (b) seven-day organ cultures of the newt skin; (c) seven-day organ cultures of the newt skin treated with the acidic RP with pI <3.0 at a concentration of 1×10^{-11} mg/ml; (d) fourteen-day organ cultures of newt skin; (e) fourteen-day organ cultures of newt skin treated with weakly acidic RP with pI 4.5–5.1 at a concentration of 1×10^{-13} mg/ml; (f) fourteen-day organ cultures of newt skin treated with basic RP with pI 8.5–9.5 at a concentration of 1×10^{-12} mg/ml. Magnification: ocular 10×, objective lens 20×.

On day 14, the protective effect of serum RPs was manifested even stronger. Figure 2d shows a control culture in which significant degradation of the skin can be observed: the structure of the epithelium and dermis is severely disturbed, and the dermis abounds in dead cells. By contrast, the cultures treated with serum RPs contained functional glands that continued to secrete mucus, as the stratified epithelium had an abundant mucous coating. Cell degradation in the dermis was distinct than in the control, and pigmented cells at the epidermis–dermis boundary formed a dense layer (Figures 2e, 2f).

It should be noted that intensification of mucus secretion, renewal of skin epithelium, and increase in the number of pigmented cells take place when amphibians molt or are exposed to some adverse external factors [Vorontzova et al., 1952].

Thus, blood serum RPs at very low concentrations proved to have a distinct protective effect on the cultures of newt skin explants.

This result was confirmed in experiments on the effect of weakly acidic serum RP (pI 4.5–5.1) on skin wound healing in mice.

In the first control group (without any treatment), almost complete wound epithelization was observed on day 11 after surgery, with the signs of inflammation being insignificant (Figure 3a). The dermis contained many capillaries and dense collagen strands extending parallel to the epidermis; i.e., the formation of a fibrous scar took place. Fat cells had an irregular shape and were separated by connective tissue. A few regenerating ducts of skin glands could be detected. Scar formation was also observed in the subcutaneous tissue, where collagen fibers were less dense than in the dermis and fat cells were almost absent.

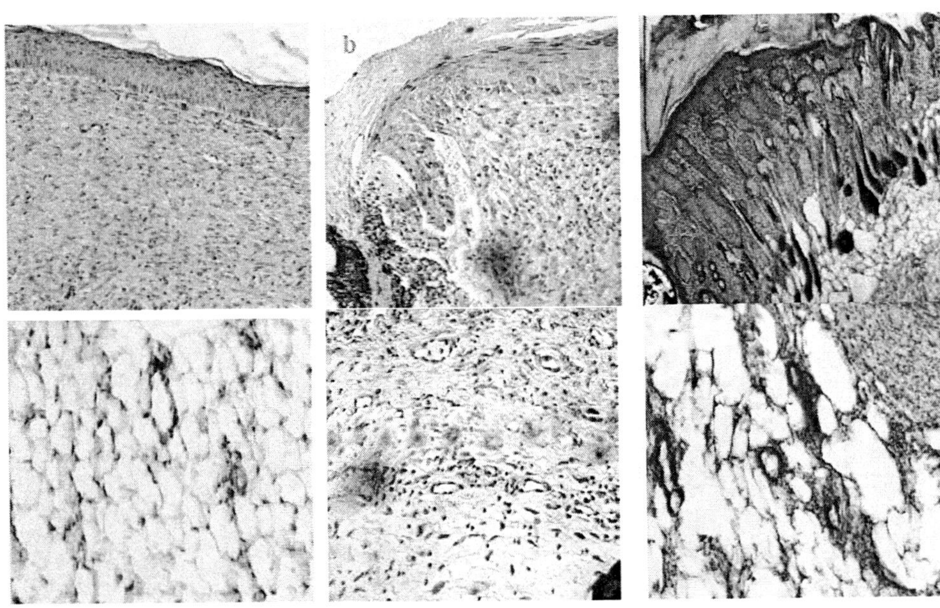

Figure 3. Transverse section of the wound healing murine skin (a) first control group, the wound remained untreated; (b) second control group, the wound was irrigated once a day with 100 □l of a saline; (c) the experimental group, the wound was treated once a day with 100 □l of a solution containing weakly acidic RP (pI 4.5-5.1) at a concentration of 1×10^{-10} mg/ml. Magnification: ocular 10×, objective lens 20×.

In the second control group (irrigation with a saline), epithelization of the wound on day 11 was incomplete, a focus of chronic inflammation appeared in its central part, and a connective tissue scar formed in the dermis (Figure 3b). Apparently, wound treatment with a saline provided favorable conditions for the development of pathogenic microorganisms, and, hence, inflammation in these animals was more severe than in the first control group.

The picture of wound healing in the third, experimental group was markedly different: wound edges strongly converged, and epithelization was better manifested than in the first two groups (all layers of the skin epithelium were restored). No foci of inflammation was detected. The structure of the dermis was almost normal. Numerous gland ducts and hair follicles were restored in the dermis and subcutaneous tissue, adipose tissue developed, and circular muscles regenerated in the wound region of skin (Figure 3c). Therefore, the weakly acidic serum RP effectively stimulated wound healing in mice, providing for rapid restoration of morphologically normal tissue in place of skin wound.

Thus, the weakly acidic serum RP (pI 4.5–5.1) was localized in the intercellular space of the newt liver, which is evidence that it is probably synthesized in this organ. The same RP facilitates the healing of skin wounds in mammals, stimulating the processes of tissue regeneration and repair. All three serum RP fractions have a protective effect on organ cultures of the newt skin.

REFERENCES

E.I. Bueverova, E.V.Bragina, M.M.Reznikova, V.P. Yamskova, N.G. Khrushchov, *Doklady Akademii Nauk,* Vol.281, N 1, pp. 158-160, 1985.

E.N.Grigoryan, M.S. Krasnov, K.S. Aleynikova, V.A. Poplinskaya. V.I. Mitashov, *Doklady Akademii Nauk,* Vol.404, N 5, pp. 566-570, 2005.

R.A. Gundorova., I.P. Khoroshilova-Maslova., E.V.Chentsova, L.V.Ilatovskaya, V.P.Yamskova, and I.Yu. Romanova, *Vestnik Oftal'mologii,* Vol. 113, N 2, pp. 12–15, 1997.

N.V.Engelhardt "Immunokhimicheskii analiz" Moscow, *Nauka*, pp. 165–188, 1968.

N.V. Engelhardt, "Synthesis of blood plasma proteins and alpha-fetoprotein in ontogeny", pp. 92–100, in the book "Immunologicheskie aspekty biologii razvitiya", Moscow, *Nauka*, 1984, 346 pp.

E. Engvall, P. Perlman, *Immunochemistry*, Vol. 8, N. 9, pp. 871–874, 1971.

J.P. Gosling, *"Immunoassays"*, pp. 19–36, A Practical Approach", Ed. by J.P. Gosling, New York, Oxford University Press, 2000, 215 pp.

B.D Hames, "One-dimensional gel electrophoresis of proteins", pp. 1-44, in the book *"Gel Electrophoresis of Proteins. A Practical Approach"*, Ed. by B.D. Hames and D. Rickwood, New York, Oxford University Press, 1990, 106 pp.

M.S.Krasnov, E.N. Grigoryan, V.P. Yamskova, Izvestia Rossiyskoy Akadimii Nauk, *Seria Biologichiskay*, N. 1, pp. 22–36, 2003.

A.T. Mikhailov, V.N. Simirsky, "Immunochemical Methods in Developmental Biology", Moscow, *Nauka,* 1991, 288 pp.

P.A. Riley, *Int. J. Biochem. Cell Biol.*, Vol. 29, pp.1235-1239, 1997.

G.Sichel, M.Scalina, C.Corcoro, *Microsc. Res. Techn.,* Vol. 57, pp. 477–490, 2002.

P.K. Smith, R.I. Krohn, G.T. Hermanson, A.K. Mallia, F.H. Gartner, M.D. Provenzano, E.K. Fujimoto, N.M. Goeke, B.J. Olson, D.C. Klenk, *Analitical Biochemistry*, Vol. 150, N 1, pp. 76-85, 1985.

M.A. Vorontzova,, L.D. Liozner, I.V. Markelova, E.Ch. Pukhal'skaya, "Newt and Axlotl", Ed. by M.A. Vorontzova, Moscow, *Sovetskaya nauka,* 1952, 296 pp.

I.A Yamskov., A.A Vinogradov., A.N Danilenko., L.A. Maslova, V.P. Yamskova, Prikladnaya. *Biokhimiya i Mikrobiologiya,* Vol. 37, N 1, pp. 36–42, 2001.

V.P. Yamskova, M.M Reznikova, *Zhurnal Obshchey Biolologii,* Vol. 52, N 2, pp. 181–191, 1991

V.P.Yamskova, E.A Modyanova, V.I. Leventhal, T.P. Lankovskaya, O.K. Bocharova, A.G. Malenkov, *Biofizika,* Vol. 22, pp. 168–174, 1977a.

V.P.Yamskova, E.A. Modyanova, M.M. Reznikova, A.G. Malenkov, *Molekulyarnaya Biologiya,* Vol. 11, N 5, pp. 1147–1154, 1977b.

V.P .Yamskova, N.V .Nechaeva, N.B .Tumanova, Yu.G. Yurovitskii, T.E. Novikova, V.I. Fateeva, I.G. Gvazava, Izvestia Rossiyskoy Akadimii Nauk, *Seria Biologichiskay,* N 2, pp. 190–196, 1994.

V.P. Yamskova, N.B. Tumanova, A.S. Loginov, *Byulleten' eksperimental'noy biologii I meditsiny,* N.3. pp.3O3-3O6, 1990.

In: Biochemical Physics Frontal Research
Editors: Sergei D. Varfolomeev et al., pp. 79-126
ISBN 1-60021-425-8

Chapter 9

NEW APPROACHES TO DETERMINE HUMAN BIOLOGICAL AGE

A. N. Strelnikov and V. B. Mamaev
N. M.Emanuel Institute of Biochemical Phisics,
Russian Academy of Sciences

ABSTRACT

Age-related changes of 20 systems and organs with men and women have been studied and their reliability estimated. Diagrams and tables reflecting those age-related changes are represented. Mathematic formulas determining biological age are cited for every particular organism system allowing for sex. Experimental data have been obtained with the help of electropunctual diagnostics method.

Keywords: Biological age, ageing, electropunctual diagnostics

INTRODUCTION

Nowadays one of new lines in gerontology and practical medicine enjoys active and dynamic development – aging prophylaxis. A network of Antiaging Centers appeared in many countries. In these conditions it has become increasingly important to create a gerontological method of determining biological age (BA) as a method of measuring organism aging. However specialists working in this field face a number of unsolved problems.

Table 1. Human organism architectonics and its reflection in the measured points

Apparatus	Systems	Organs and subsystems.	CSP
Metabolic apparatus	Digestive	*Alimentary tract*: oral cavity, pharynx, esophagus, stomach, small intestine, large intestine.	E44b, E3, RPd1a, IG1b, F1a, VB43b, GI1b
		Alimentary tract glands: salivary glands, liver and gallbladder, pancreas, small glands of mucous membrane of alimentary tract parts	
	Respiratory	*Air-conducting organs*: upper air passages, bronchi, trachea	P10c, P10,9
		Gas exchange organs: lungs	
	Cardiovascular	*Blood circulation:* heart, arteries, veins	MC8d, C8c, AL1c, LI11
		Lymphatic: lymphatic vessels and nodes of separate body parts	
	Excretory	Kidneys, urinary bladder, ureters, urethra, skin sudoriferous glands, excretory gland of digestive system.	R1-3, V67, V66b, V65
Protective apparatus	Antioxidation	Enzymatic antioxidation system, vitamin circuits, etc. metabolite functioning as antioxidants	AL1b, F1a
	Immune	Marrow, thymus, lymphoid tissue of respiratory and digestive system walls, lymph nodes, spleen	RPs1a, AL1b, PAD1b, E11, LI4, LI12
	Barrier	Skin, epithelial tissue of respiratory and digestive systems, large intestine mucous membrane	AL1, AL2, K1-3, PAD1b
	Neuro-endocrine	*Limbic*: hypophysis, epiphysis and other endocrine glands	TR1b, VB12, VB21, V8
Regulatory apparatus	Nervous	*Central nervous*: brain and spinal marrow.	ND1b, VB7, VB9, VB20, T13
		Peripheral nervous: cranial and spinal nerves	
		Vegetative nervous: sympathetic and parasympathetic parts	
	Endocrine	*Endocrine glands*: epiphysis, hypophysis, neurosecretory nucleus of brain hypothalamic branch, thyroid and parathyroid glands, adrenal glands.	TR1b, TR20, E10
		Mixed glands: thymus gland (thymus), pancreas, testicles, ovary.	E11, RPd1a, V65
	Analyzer	*Organs of sense*: organs of taction, eyesight, olfaction, audition, equilibrium, taste	TR17, TR18, TR21, VB1, E4
Locomotor apparatus	Bone	Skull bones, thorax bones, backbone, appendicular skeleton	V12, V11
	Vascular	Articulations and ligaments	SD1b, VB33; TR15
	Muscular	Muscles, tendons	VB34, IG9
Reproduction apparatus	Endocrine	Hypophysis, sexual glands, prostata gland	TR1b, TR3, TR1
	Reproductive	*Male reproductive organs*: testicles with appendages, deferent duct, seminal vesicle, prostate and bulbourethral glands, penis, scrotum;	V66b, V65, V64, F11, E17
		Female reproductive organs: ovary, uterine tube, uterus, vagina, mammary gland	

The first of them is connected with the fact that study of age-specific change processes in the organism has not detected any single reason and single aging mechanism, on the contrary, there is a wide spread view on multiple reasons for of these processes. All organs and systems undergoing age changes influence each other and thus, make up a peculiar picture of aging every time. Moreover the speed of age-specific change is fully individual and depends on both genetic and environmental peculiarities. Consequently, when determining BA, there arises a necessity to take into account the state of all organism systems.

The performed analysis of existing BA determination methods has revealed that there are no methods covering all systems of the organism.

Below we are giving a list of all human systems, subsystems and organs (Table 1). Te table indicates what systems of the organism are to be taken into account when developing the BA determination method.

The second difficulty is that approaches and methods which are successfully applied to decode molecular-biological life mechanisms in recent decades are practically useless in this case. In addition to intricate problem the case becomes complicated because we cannot use tissues, intercellular fluid and other structures of human internal environment for researches without harm to health. Therefore, it is necessary to used noninvasive methods to determine BA.

Besides, the method should have cumulative character, i.e. should be able to integrate the information about the state of certain subsystems.

From our viewpoint, a method of electropunctual diagnostics with a possibility to cover all systems of the organism without infringing their integrity is promising for this purpose [1].

In the right column of Table 1 measurement points used in the electric puncture method to research the related systems and organs are shown. The work uses abbreviations as those customary in literature [1]: CSP – control measurement points. Points legend is presented in clause 2.1.

1. Materials and Methods

Electropunctual method is being successfully used by many doctors in the field of functional diagnostics during the recent decades. Method essence lies in measuring electrical conductivity of skin integument zones being the projection of biologically active points (BAP). On the basis of measurement results analysis and comparison, the conclusion determining what organs and systems interconnected with BAP have stronger pronounced abnormalities of adaptive regulation or are involved in the pathologic process.

Advantages of the method: high information value, non-invasive character, speed of the procedure, comfort and convenience for the person examined, simplicity for the experimentalist, cheap equipment.

During the examination measuring current 5.5 – 11.25 mcA with maximum voltage 1.2 V [2] is used. When measuring BAP electrical parameters, maximum deflection denotes BAP measuring level (ML) in the electropunctual diagnostics.

Below there is a table interpreting measurement results of BAP electric parameters (see Table 2).

With the help of this method 415 women and 170 men living mainly on the territory of Moscow and Moscow Region were examined. The whole contingent both men and women was divided into 9 age groups (see Table3).

Table 2. Interpretation of instrumental indexes

ML values (c.u.)	Interpretation
90 – 100	Acute or catarrhal inflammatory process
86 – 89	Subacute, focal or local inflammatory process
66 – 85	Hyperfunctioning, toxic load
50 – 65	Norm
40 – 49	Hypofunctioning, lassitude
30 – 39	Initial degeneration
21 – 29	Degeneration, partial cellular structure atrophy of an organ or tissue system
0 – 20	Total atrophy or neoplastic transformation of cellular structures of an organ and tissue system

Table 3. Contingent age structure

Age group (Δx)	Women		Men	
	Number of patients (n)	Average age ($\bar{x}$)	Number of patients (n)	Average age ($\bar{x}$)
1-4	3	3.67	4	3.25
5-14	19	9.63	26	9.38
15-24	35	20.09	22	20.45
25-34	63	29.95	19	30.63
35-44	112	40.07	32	39.56
45-54	92	49.53	35	49.51
55-64	67	59.19	22	59.05
65-74	21	70.10	9	67.56
75-84	3	76	1	75
Σ	415		170	

2. Results and Discussion

2.1. Age-Related Changes of Organism Systems

We have studied age-related changes of the measuring level value in the control points of related systems or organs and estimated their credibility.

The following systems and organs were examined: lymphoid system (LI), respiratory system (P), large intestine (GI), nervous system (ND), vascular system (MC), immune system -1 (AL), immune system-2 (PAD), endocrine system (TR), heart (C), small intestine (IG), immune system -3 (RPs), pancreas (RPd), liver (F), articulations (SD), stomach (E), connective tissue (STD), skin (K), adipose tissue (GD), gallbladder (VB), kidneys (R), urogenital system (V).

On Pictures 1÷42 one can find figures reflecting age-related changes of the measuring level value wit women and men for every system being examined. Along the abscissa axis an average age of the examined groups is plotted, and along the ordinate axis the calculated average values of the measured figures are plotted in conventional units (c.u.) for every age group.

A black full line connects the measuring level figures of the right part of the body and a red full line connects the measuring level figures of the left part of the body. For figures of the measured values with the related age there are confidence intervals (CI) plotted on the figures. Black and red dotted lines indicate regression lines for the right and left parts of the boy correspondingly.

2.1.1. Lymphoid System

Lymphoid system control measurement point sums up the information about 24 system parameters including the information about tonsil, lymph outflow from ear, Pirogoff-Waldeier lympho-epithelial ring of pharynx, tubal tonsil, lymph outflow from upper and lower jaw, lymph outflow from mucous coats of nasal cavity and accessory sinuses of the nose, lymphatic vessels and pulmonary nodes [1, 2].

On pictures 1÷2 there are figures reflecting age-related changes of measuring level value in the control lymphoid system point on the right and left sides of female and male body. Statistic values of measuring level for all age groups lie within the range of physiological norm (50÷65 c.u.). With men the inclination of regression line for the right side is unreliable (bd=-0,08±0,07, where bd – regression coefficient).

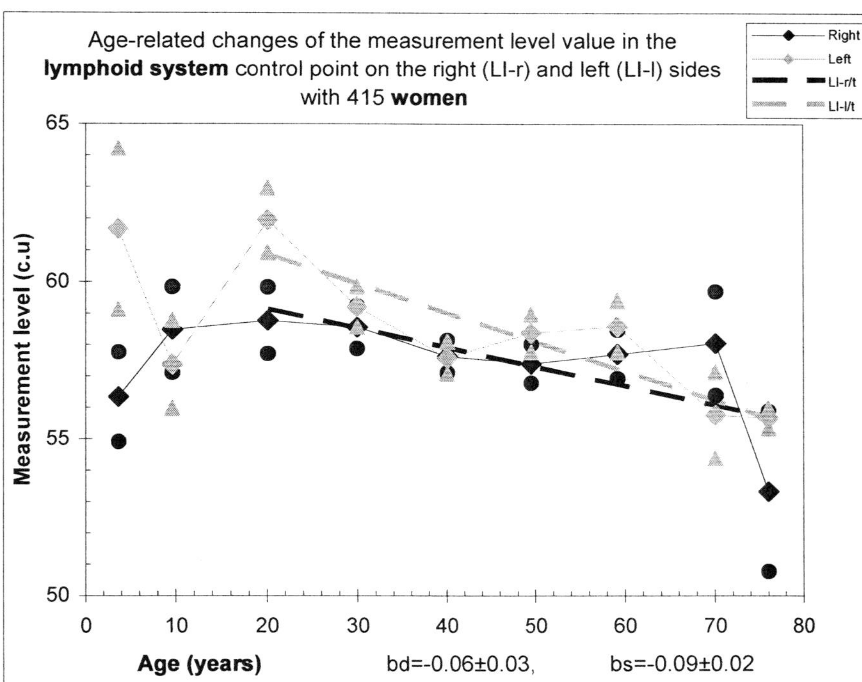

Picture 1.

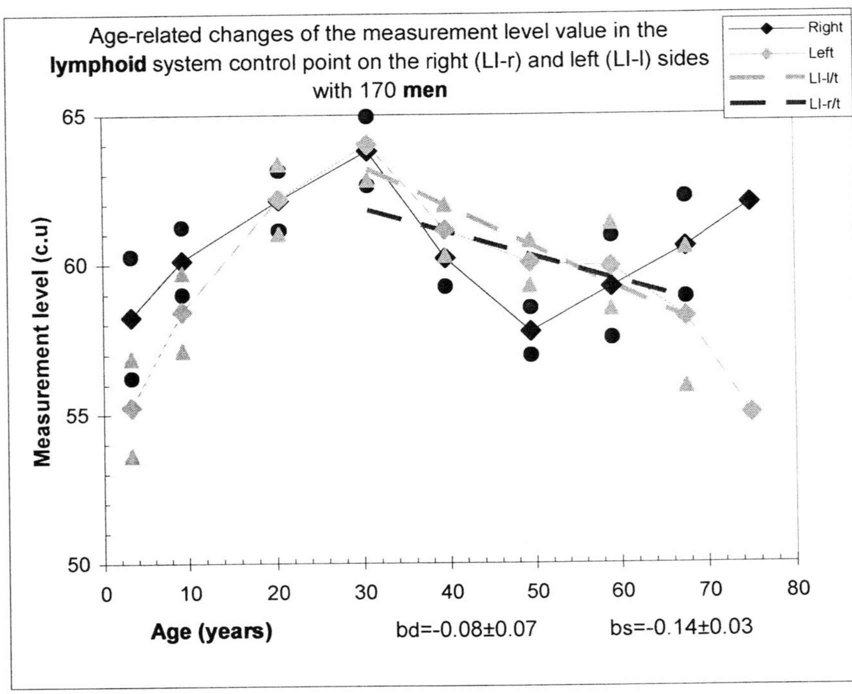

Picture 2.

With women a rapid uprise from 20,09 years old to 61,94±1,04 c.u. and fall at the age of 76 to 55,67±0,33 c.u. for the left side are reliable of all age-related changes of the measurement level. Whereas changes of all other value are unreliable (see Picture 1).

As regards to men, there is a reliable maximum on the right and left sides at the age of 30,63 amounting at 63,79±1,18 c.u. and 64,0±1,2 c.u. correspondingly (see Picture 2.)

2.1.2. Respiratory System

The respiratory system control measurement point provides the information about operation of 40 related organs and tissue structures including lungs, teeth ridge, bronchial tubes, visceral pleura layer, bronchi, trachea [1, 2].

Reliable maximum of measurement level with women is fixed at the age of 20,09 for both sides and equals to 59,06±1,23 c.u. and 59,06±1,01 c.u. correspondingly, while a reliable minimum is distinguished at the age of 76 on the left side and equals to 50,0±2,08 c.u. (see Picture 3.).

Men's maximum value of measurement level is reached at the age of 30,63 and equals to 62,16±1,26 c.u. for the right side, and 63,32±1,36 c.u. for the left side. Minimum values of measurement level are revealed at the age of 75 and lie below the physiological standard corridor for both sides - 47 c.u. for the right side and 49 c.u. for the left side (see Picture 4.). With men the inclination of regression line for the right side is unreliable (bd,=-0,11±0,09).

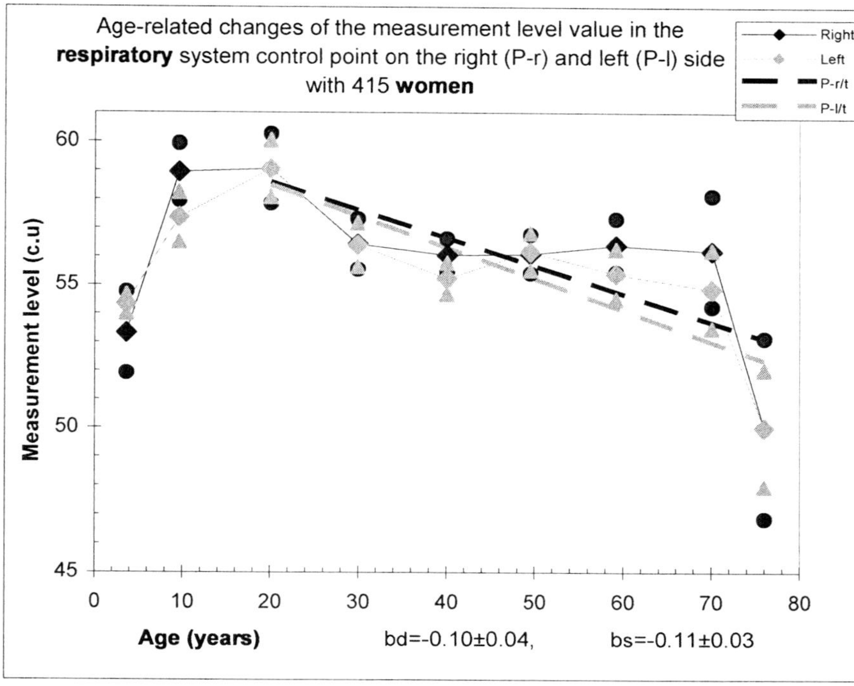

Picture 3.

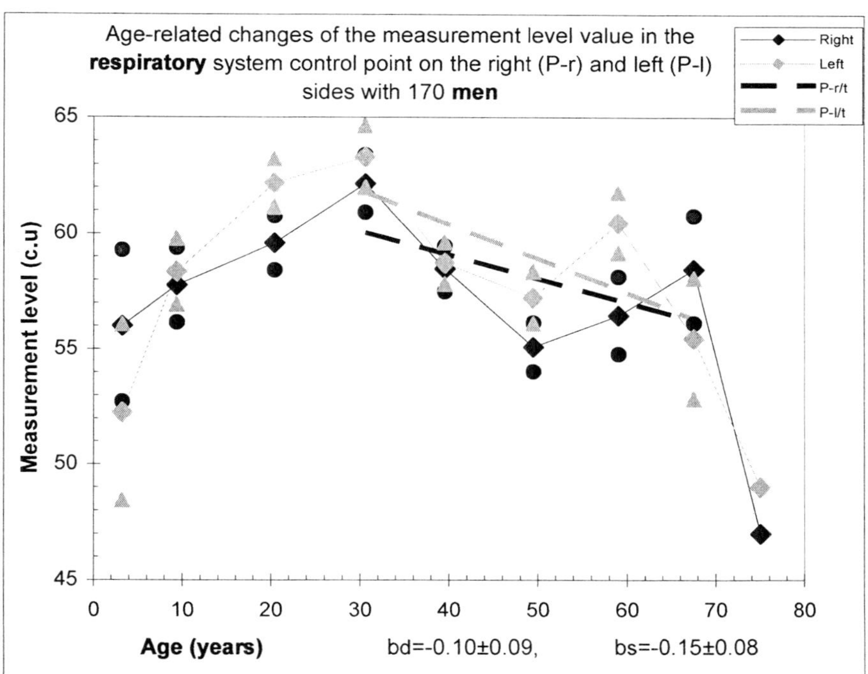

Picture 4.

2.1.3. Large Intestine

The large intestine control measurement point carries information about 64 parameters having informational and functional correlation with different parts of large intestine. Among them there is transverse colon, sigmoid colon, large intestine peritoneum, left and right large intestine flexure, descending and ascending parts of large intestine, blind gut, appendix, colic lymph nodes [1, 2].

With women large intestine age-related changes are reflected on Picture 5. Maximum value of measurement level (ML) is observed at the age of 9,63 and equals to 50,74±1,17 c.u. for the right side and 51,00±1,11 c.u. for the left one. Starting from the age of 20,09 ML values remain below the physiological standard corridor and decrease to 41,00±1,53 c.u. for the right and to 41,00±2,08 c.u. for the left side. Regression line inclination (regression coefficient) equals to -0.14±0.01 for the right side and -0.13±0.01 for the left side. All age differences in measurement level values are reliable.

Men also show the decrease of measurement level after the age of 9,38 but only changes starting from the age of 39.56 are reliable. (see Picture 6). From the age of 39,56 all ML values are located below the physiological standard corridor. Regression line inclinations equal to -0.17±0.02 for the right side and –0.16±0.03 for the left one.

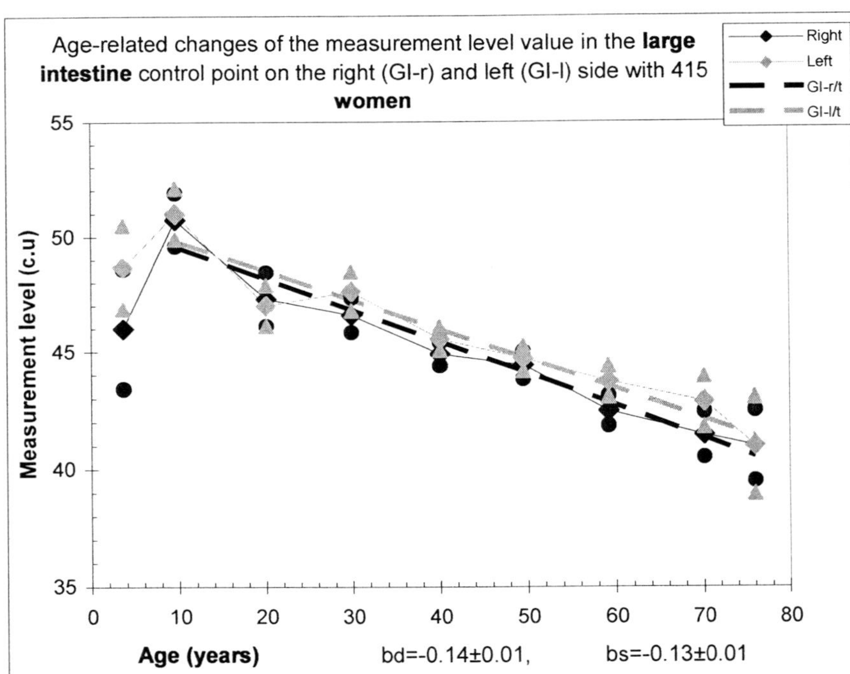

Picture 5.

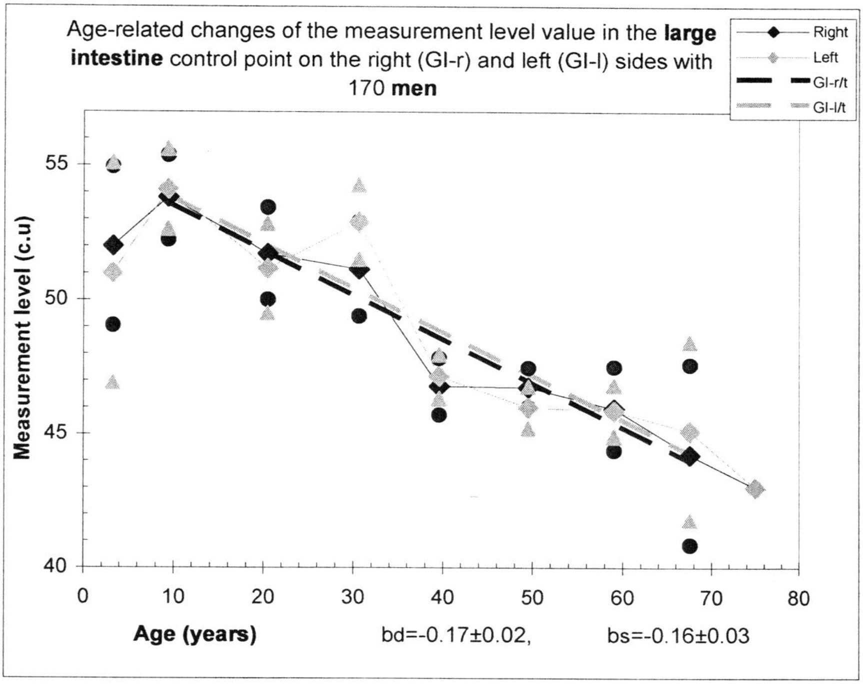

Picture 6.

2.1.4. Nervous System

The nervous system control measurement point carries information about 8 system parameters, namely about the condition of lumbar cord and sacral part of spinal cord, peripheral and central nervous system, meninges, cervical and dorsal parts of spinal cord, brainstem, cranial nerve [1, 2].

Women have the inclination of regression lines equaling to -0.12±0,03 for the right side and -0.13±0,02 for the left side. Men have correspondingly -0.16±0.02 for the right side and -0.14±0.03 – for the left side. All inclinations of regression lines are reliable.

With women starting from the age of 29.95 ML values for the right side are below the physiological standard corridor, and for the left side – they are lower for all age groups apart from the age of 9,63. Men have these values below the standard after the age of 39,56 for both sides (see Pictures 7÷8.). On each of these figures two plateaus can be distinguished. Women have the first plateau for both sides under the age of 20,09 inclusive, and the second plateau from 29,95 to 70,1 years old. With men the first plateau for both sides is observed under 30,63 years old inclusive, and the second is observed from 39,56 to 67,56 years old.

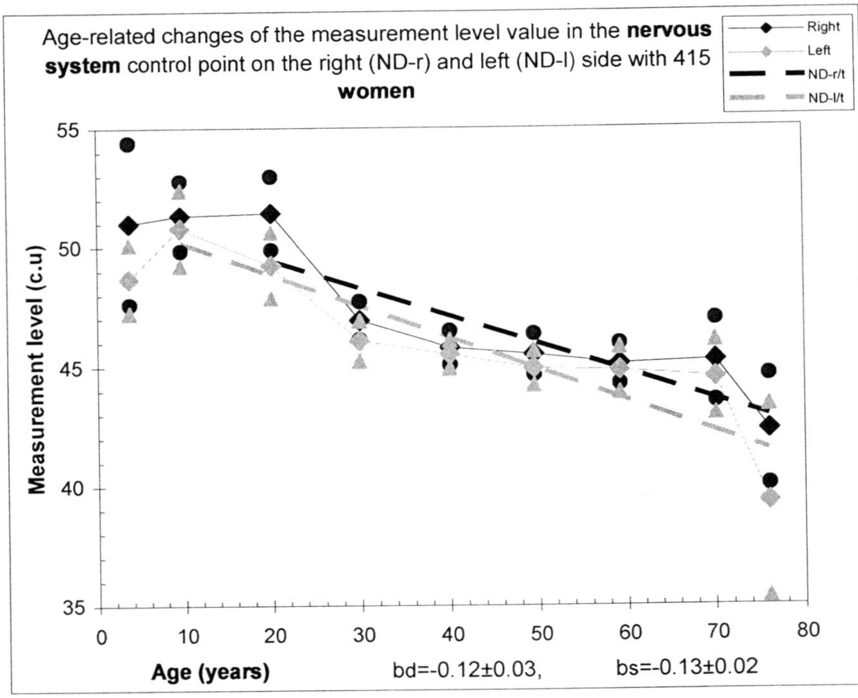

Picture 7.

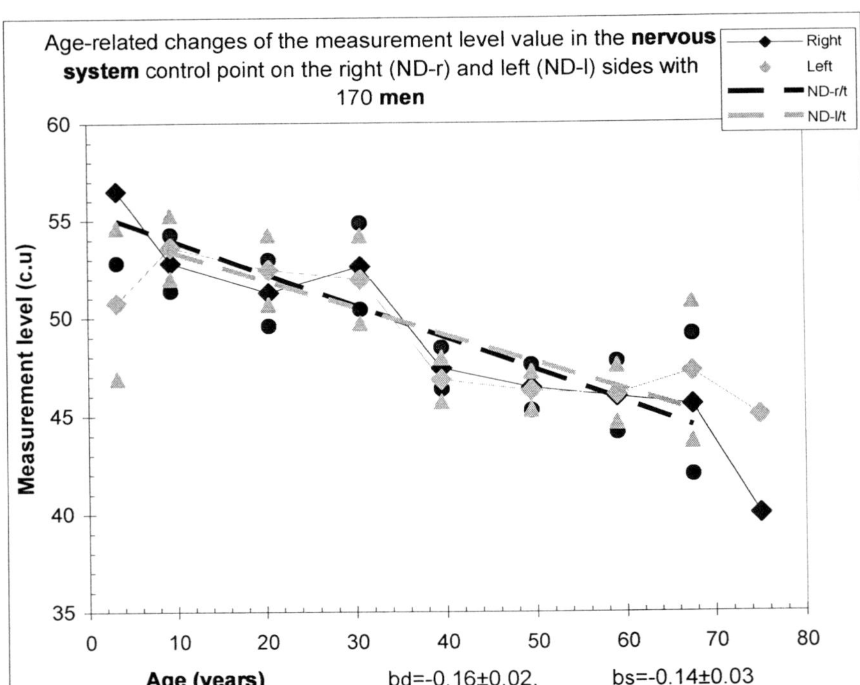

Picture 8.

2.1.5. Vascular System

The vascular system control measurement point carries information about 36 parameters including the condition of arteries, initial part of thoratic duct flow, thoratic duct flow, veins, coronary plexus, coronaria.

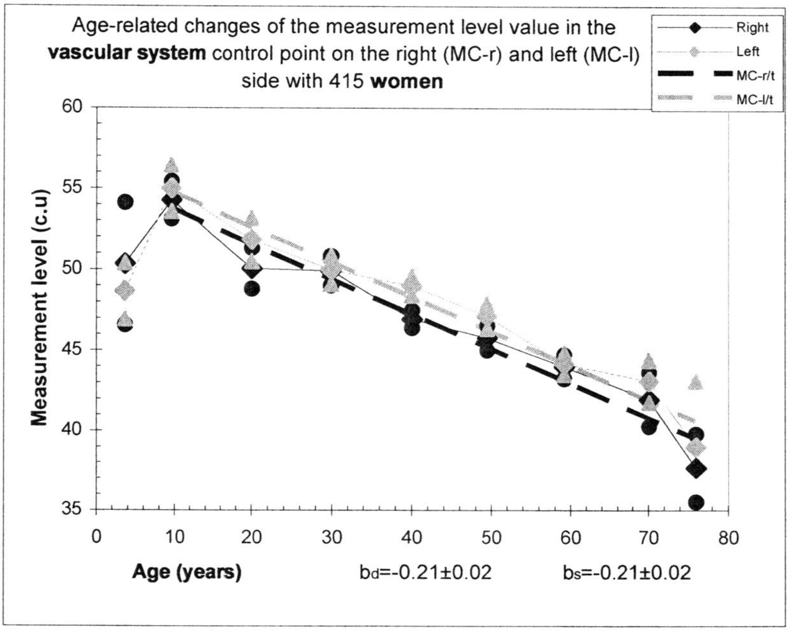

Picture 9.

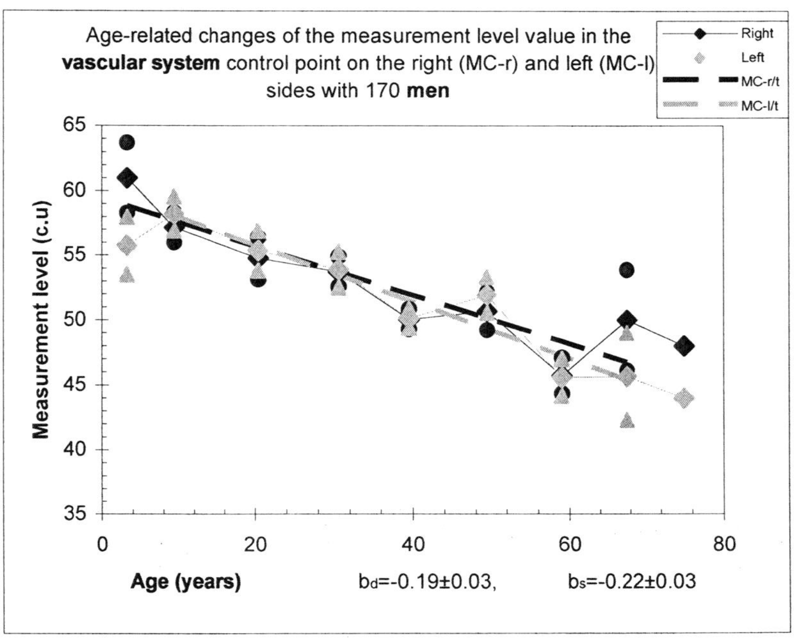

Picture 10.

Pictures 9÷10 show age-related changes of vascular system with women and men. Inclinations of regression lines of the system equal to –0,21±0,02 for women for both sides, and –0,19±0,03 for men for the right side and –0,22±0,03 for the left side. All these inclinations are reliable. ML values for women starting from the age of 29,95, and for men starting from the age of 59,05 lie below thephysiological standard limits. Age-related changes of measurement level value with men and women are reliable.

2.1.6. Immune System -1 (AL)

There are three control points which reflect the condition of immune system: immune system-1, immune system -2, immune system -3. In literature it is common practice to indicate them as allergy meridian (AL), parenchymatous degeneration meridian (PAD) and spleen meridian (RPs) correspondingly.

Control point of immune system-1 reflects the condition of 7 parameters including vascular and allergic skin lesions of lower body part and lower limbs, allergic and vascular lesions of all body, allergic and vascular lesions of nape skin, upper limbs, thoracic cavity and neck organs, allergic and vascular lesions of skin and head organs [1, 2].

With women the measurement level value decreases nearly linewise from the age of 20,09 for the right side and from the age of 3,67 for the left side without any evident deflection (Picture 11). Unlike women, men enjoy the formative period till the age of 30,63, after that measurement level slowdown being close to regression lines is observed (Picture 12).

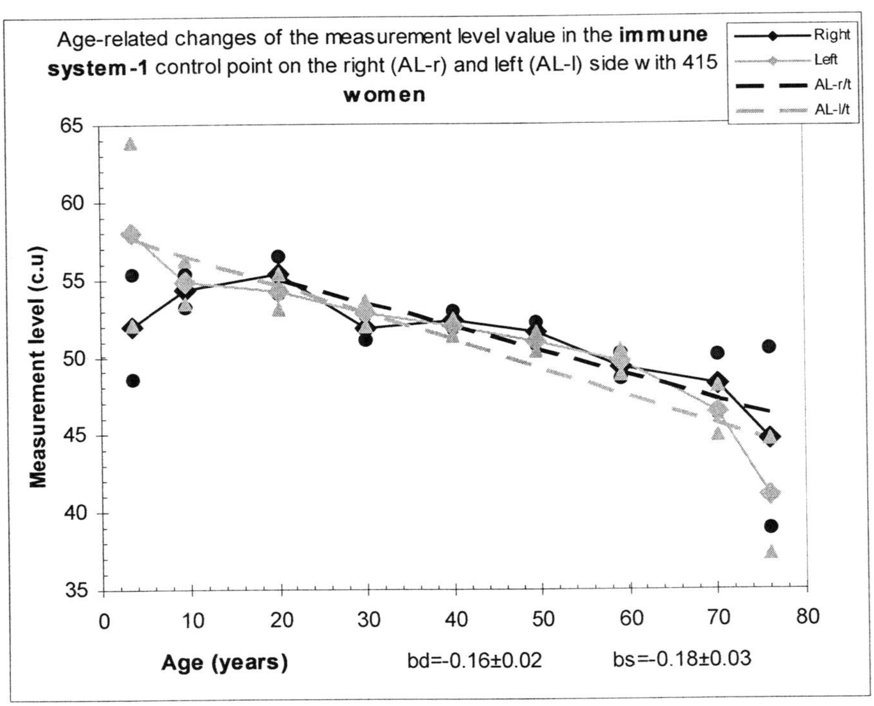

Picture 11.

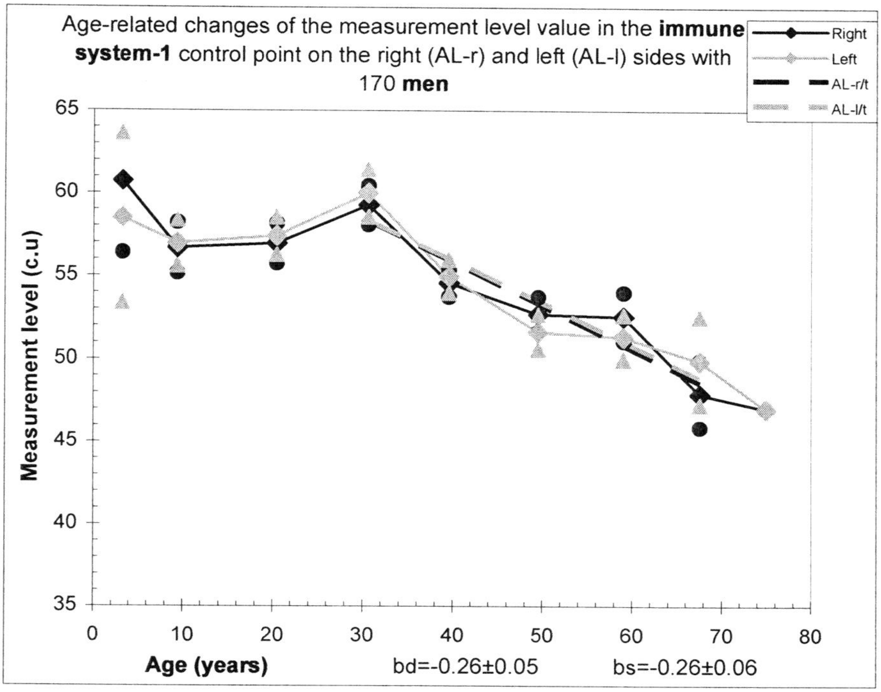

Picture 12.

Measurement level value in the immune system-1 control point on both sides of the body is reliably decreasing subject to advancing age with men as well as women. ML value remain below the physiological standard from the age of 59.19 with respect to women and 67,56 with respect to men. Inclination of regression line with women for the right side of the body equals to –0,16±0,02, and for the left side it is –0,.18±0,03. For men these inclinations equal to –0,26±0,05 for both sides.

2.1.7. Immune System -2 (PAD)

Control point of immune system -2 or epithelium degeneration and organs parenchyma meridian sums up the information about 11 points responsible for degeneration of different organs and tissues including the organs of abdominal cavity and small pelvis, peritoneum, pleura, thoracic cavity and neck, head [1, 2].

Age-related changes for women and men are represented on Pic. 13÷14. With women the inclination of regression line for the right side of the body equals to –0,14±0,01 and -0.12±0,01 for the left side. With men it is –0,24±0,06 for the right side and –0,22±0,04 for the left side. All age-related changes of the measurement level are reliable excepting the age period till 30,63 years old with men.

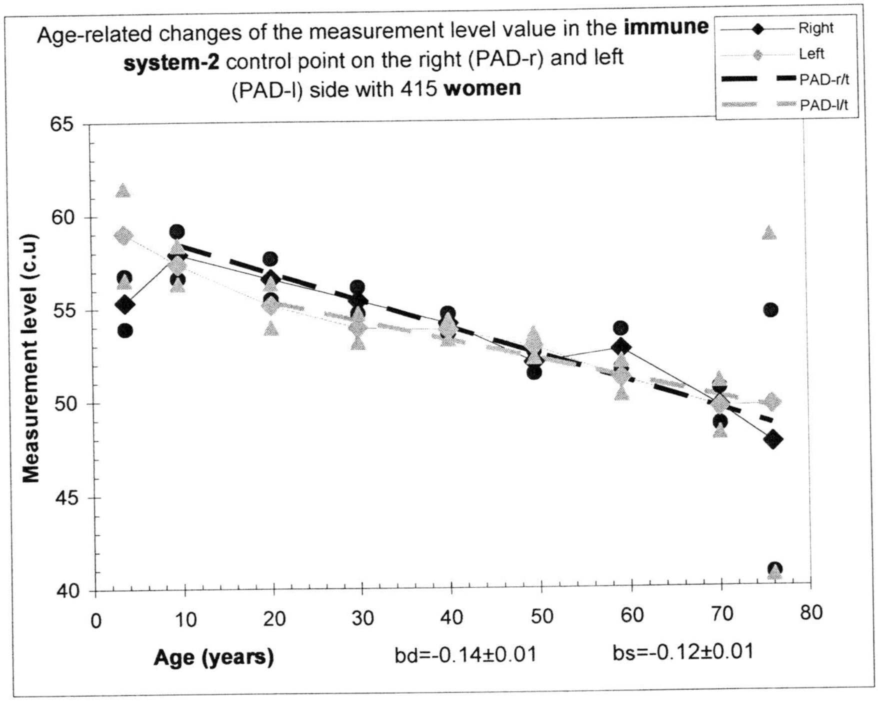

Picture 13.

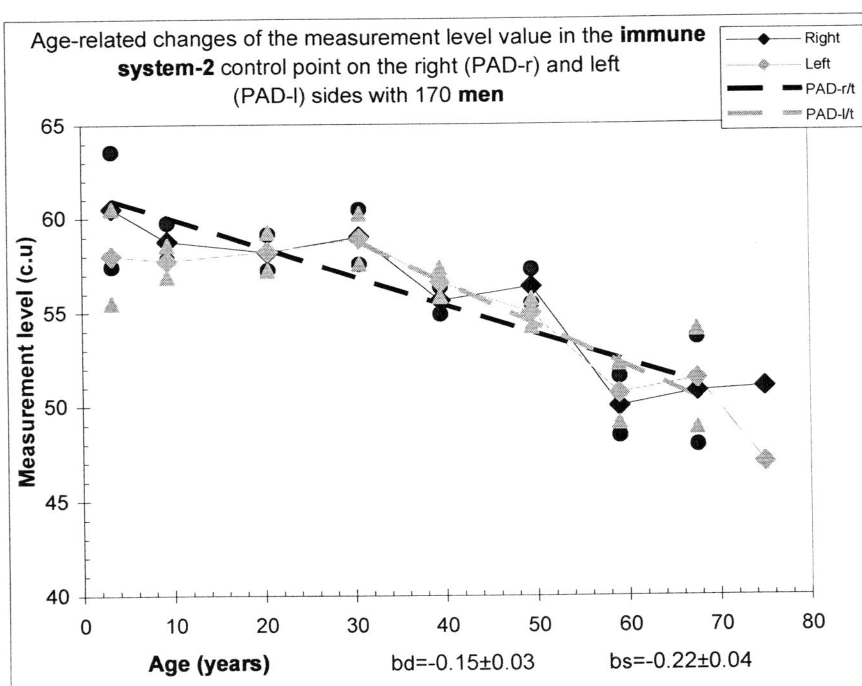

Picture 14.

2.1.8. Endocrine System

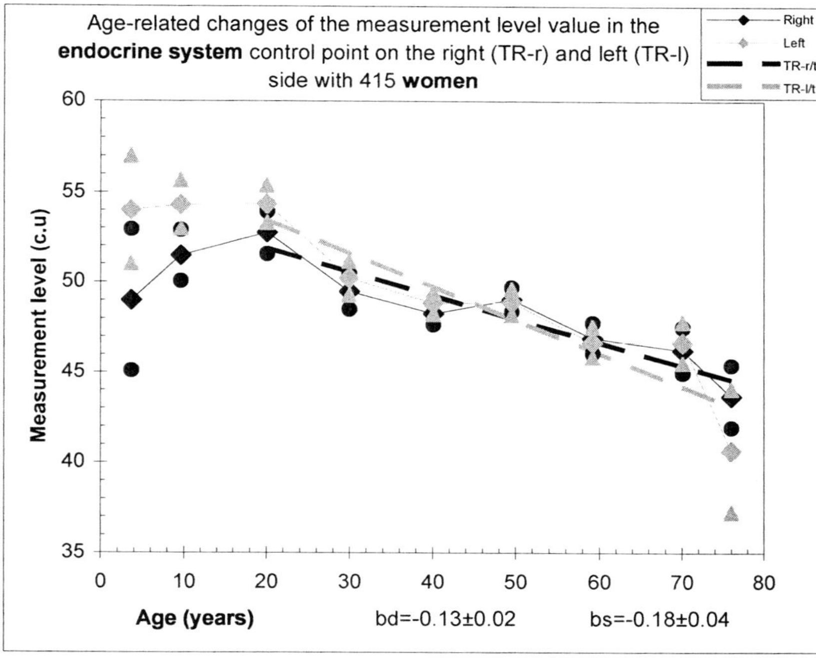

Picture 15.

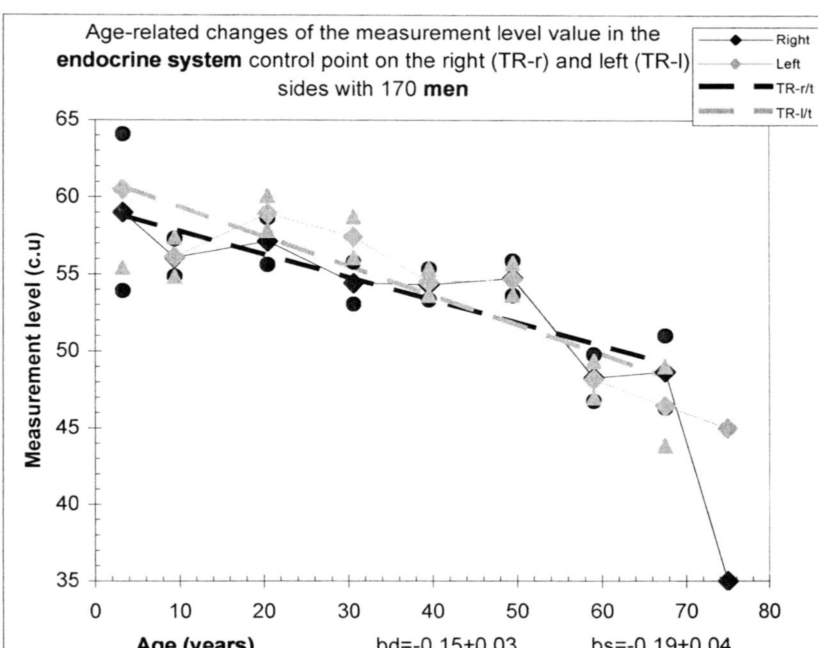

Picture 16.

When measuring the parameters of endocrine system control point we can gain the insight of general operation of its 80 components, among them there are gonads, adrenal glands, insulin and glucagon secretion, mammary glands, parathyroid and thyroid glands, thymus, hypophysis, epiphysis [1, 2].

Men and women have reliable inclinations of all regression lines for the endocrine system. For women they equal to -0,13±0,02 (right side) an -0,18±0,04 (left side), and for men -0,15±0,03 (right side) and -0,19±0,04 (left side). Reliable age-related changes are those starting from the age of 9,63 with respect to women and after the age of 59,05 with respect to men. On average, ML values lie below the physiological standard corridor after the age of 29,95 for women and after 59,05 years old for men.

2.1.9. Heart

Heart control measurement point carries the information about 36 parameters related to heart functioning. Among tem there are pulmonary valve, aortic valve, pericardium, tricuspid valve, mitral valve, atrioventricular node, left crus of His' bundle, myocardium [1, 2].

With women measurement level values in the heart control point fall nearly linewise with the increase of years and the inclination of regression is -0,14±0,03 for the right side and -0,13±0,01 for the left side. Thereat, up to the age of 20,09 women are in the progress of formation processes, after which the decrease of ML value is observed (see Picture 17.).

Another picture is observed with men. For the right and left sides a reliable inclination of linear regression equals to -0.25±0,04, and the process of development or formation lasts till the age of 30,63, further the picture is the same like that of women (see Picture 18.).

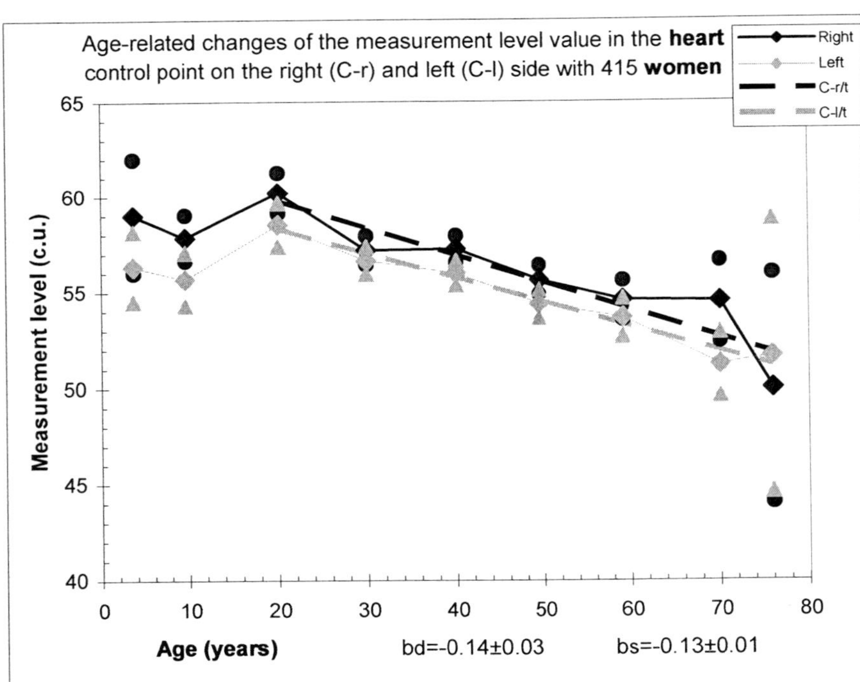

Picture 17.

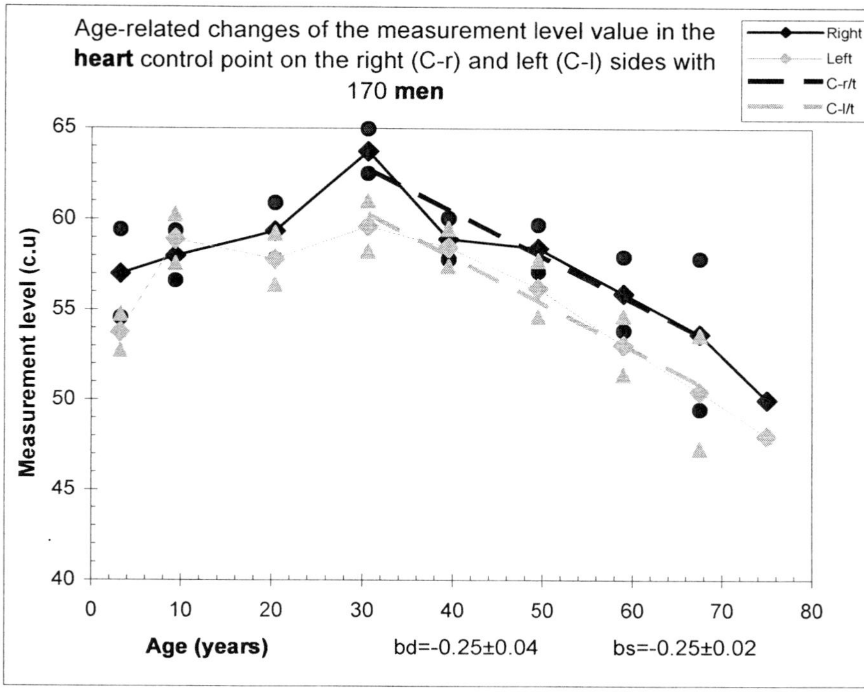

Picture 18.

2.1.10. Small Intestine

Small intestine control point carries the information about 59 representative points used in electropunctual diagnostics and therapy of alimentary tract diseases, among them there are: terminal part of ileum, duodenum, small intestine proper, duodenum peritoneum, small intestine peritoneum, lower horizontal part of duodenum, jejunum, descending part of duodenum, duodenojejunal flexure, upper horizontal part of duodenum, ascending part of duodenum.

With women a number of age-related change stages can be distinguished. The formation period lasts till the age of 9,63 when the measurement level value increase from 51,67±5,56 c.u. up to 56,21±1,35 c.u. fro the right side and from 51,00±3,71 c.u. up to 57,89±1,33 c.u. for the left side. However thses changes are unreliable. Then the value decrease and plateau establishment is observed in the age range of 29,95÷59,19 at the level of 53,12±0,99 c.u. for the right side and 54,59±1,17 c.u. for the left side. Then the measurement level values decrease (see Picture 19). Inclinations of age-related change regression lines in the small intestine control point have reliable values and equal to -0.11±0,04 and –0,13±0,03 correspondingly fro the right and left sides.

With men under the age of 30,63 the increase of measurement level values, then there is their slowdown (see Picture 20). Thereat, a reliable difference of measurement level values is observed only between the age of 9,38 and 30,63 for the left side and at the age of 59,05 and 67,56. Inclinations of age-related change regression lines in the small intestine control point have reliable values and equal to –0,20±0,04 fro the right side and -0.19±0,05 for the left side.

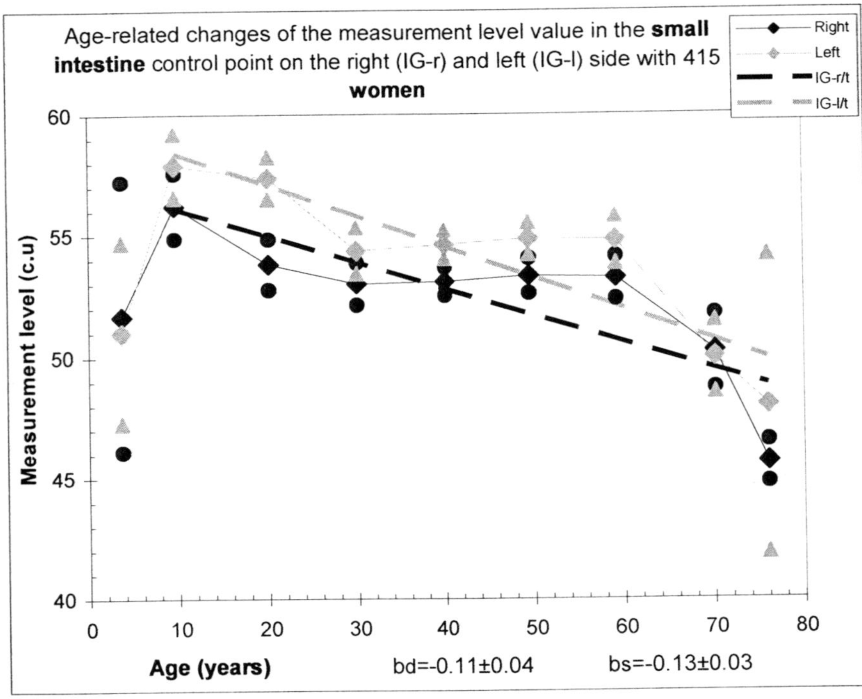

Picture 19.

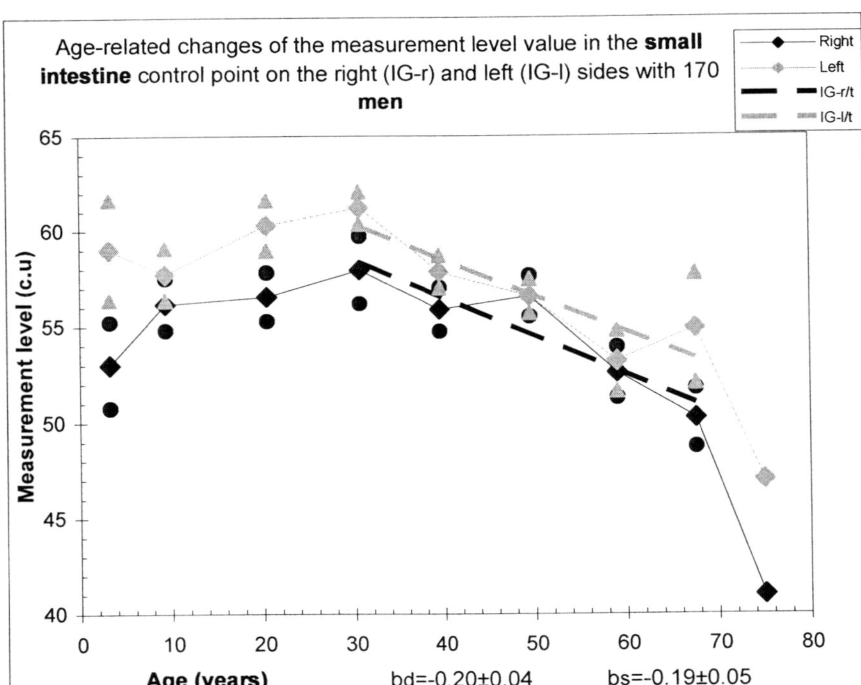

Picture 20.

2.1.11. Pancreas

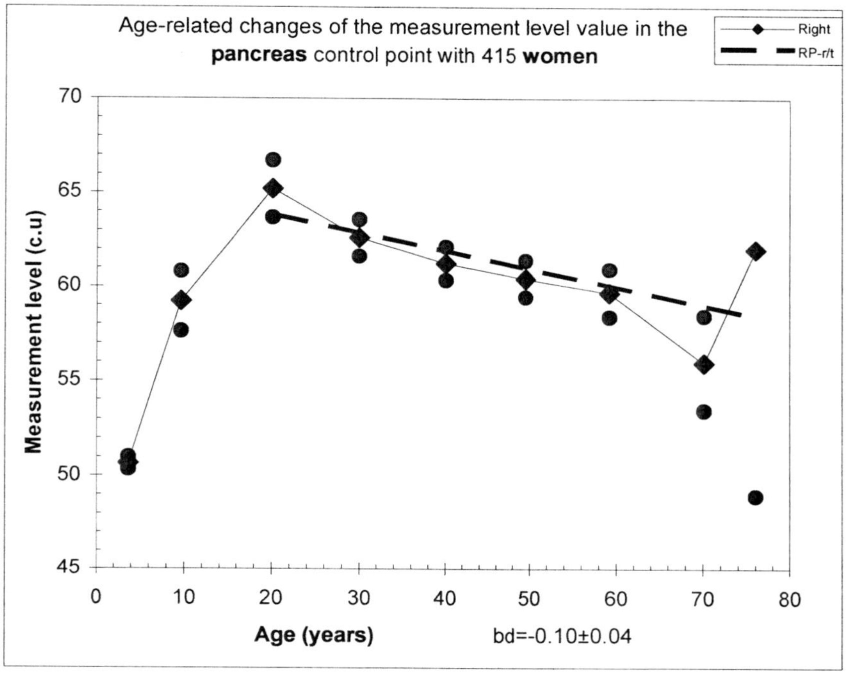

Picture 21.

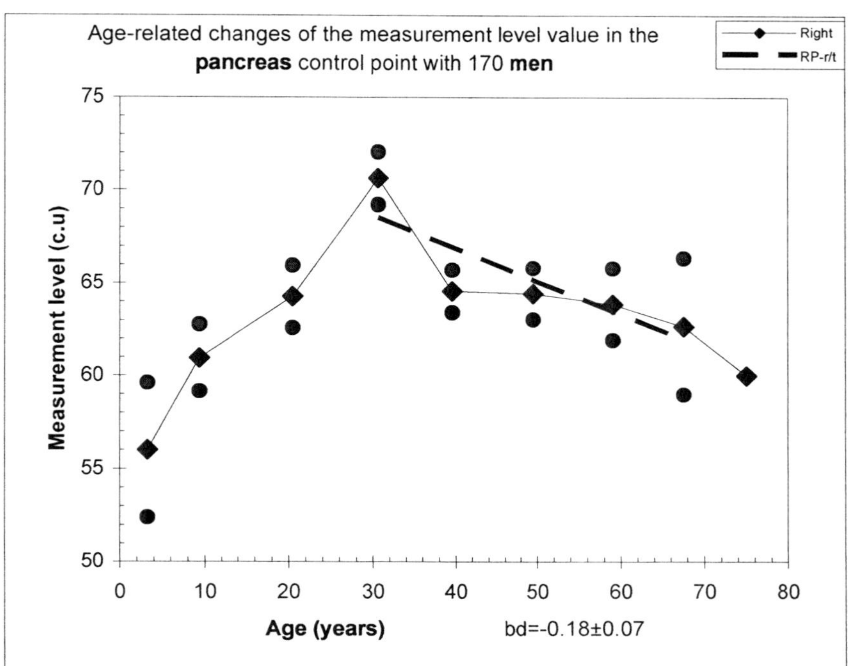

Picture 22.

When measuring electrical conductivity in the pancreas control point, we get a generalized idea of 40 parameters related to this system. Among them there are formation of protein metabolism enzymes by pancreas, pancreas proper, pancreas peritoneum, formation of nucleoprotein metabolism enzymes and uric acid metabolism by pancreas, carbohydrate metabolism enzymes, lipid exchange [1, 2].

With women maximum ML value is observed at the age of 20,09, and equals to 65,20±1,51 c.u. and minimum value at the age of 3,67 equals to 50,67±0,33 c.u. (see Pic. 21). The inclination of regression line equals to -0,10±0,04.

As for men, there is a dash at the age of 30,63 and at the same time its value exceeds the physiological standard corridor (70,63±1,45 c.u.) (see Picture 22). The inclination of regression line is –0,18±0,07.

2.1.12. Immune System -3 (RPs)

When neasuring elecrtrical conductivity in the immune system-3 control point, we get a generalized idea of 40 parameters related to this system. Among them there is the function of spleen white pulp, spleen proper, spleen peritoneum, spleen red pulp, reticuloendothelial system and reticuloendothelial network of spleen [1, 2].

With women maximum value of themeasurement level is observed at the age of 9,63 and equals to 60,05±1,53 c.u., and minimum value – at the age of 76 equals to 47,67±1,73 c.u.. Starting from the age of 59,19 the slowdown of measurement level is reliable (see Picture 23). The inclination of regression line for women equals to -0,20±0,04.

As for men, a reliable maximum of ML value was revealed at the age of 30,63 (see Picture 24). The inclination of male regression line is -0,16±0,06.

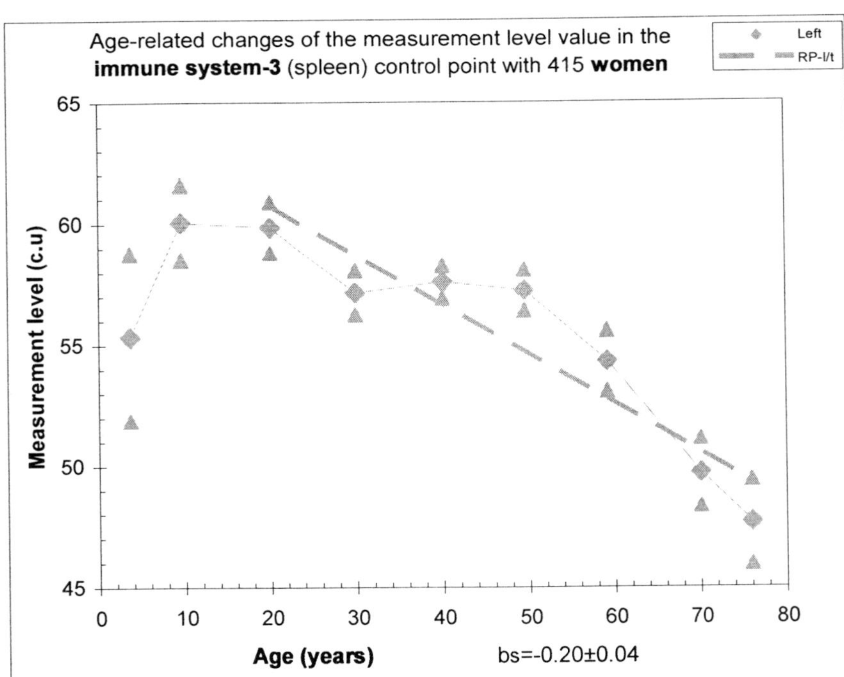

Picture 23.

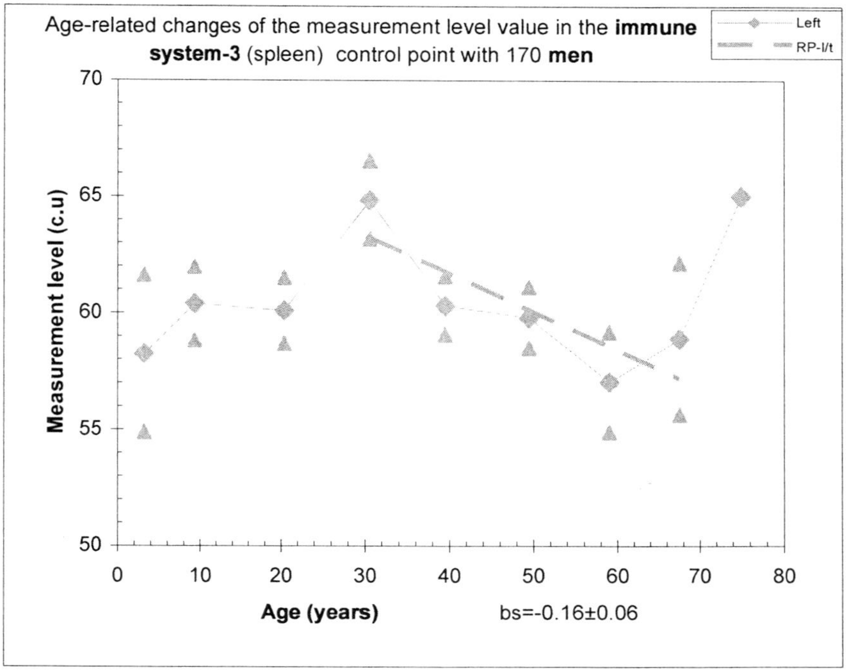

Picture 24.

2.1.13. Liver

The liver control point contains the information about 42 parameters of this organ and the related systems and tissue structures including включая central venous system of liver, liver peritoneum, liver cells and lobules, interlobular gall duct, perivascular or periportal liver system[1,2].

As for women, all points starting from the age of 20,09 lie close to the regression straight lines. ML values change from this age from 60,89±1,92 c.u. to 53,33±11,60 c.u. for the right side and from 61,79±1,55 y.e to 53,67±9,49 c.u. for the left side (Picture 25). ML regression lines in the liver control point have the inclinations equal to -0.13 ± 0,03 fro the right side and –0,12 ±0,02 for the left side.

Men has no reliable age-related differences excepting the minimum at the ages of 49,51 and 59,05 for both sides (see Picture 26).

2.1.14. Articulations

The articulations control point carries the information about the condition of 7 structural elements, namely about the condition of pelvic girdle and lower limb articulations, backbone articulations, I – II cervical vertebras and temporo mandibular joint.

As one can see from the figure (Picture 27.), it is possible to discover several stages of age-related changes. Till the age of 20,09 measurement level grow from 58,33±1,82 c.u. for the right side and from 59,67±3,85 c.u. for the left side up to 66,26±1,27 c.u. is observed. Further, there is a slowdown to 61,40±0,92 c.u. for the right side and to 61,17±0,82 c.u. for the left side till the age of 40,07. There is a plateau observed for the right side within the age range of 40,07÷59,19, and for the left side – there is a small slowdown to 58,64±1,47 c.u..

After that an evident and nearly linear drop of measurement level to 45,33±1,18 c.u. for the right side and to 41,00±2,99 c.u. for the left side is observed at the age of 76. Almost all differences in the values of measurement level are reliable. The ML regression lines for women have the inclinations of -0,36±0,07 for the right side and -0,42±0,07 for the left side.

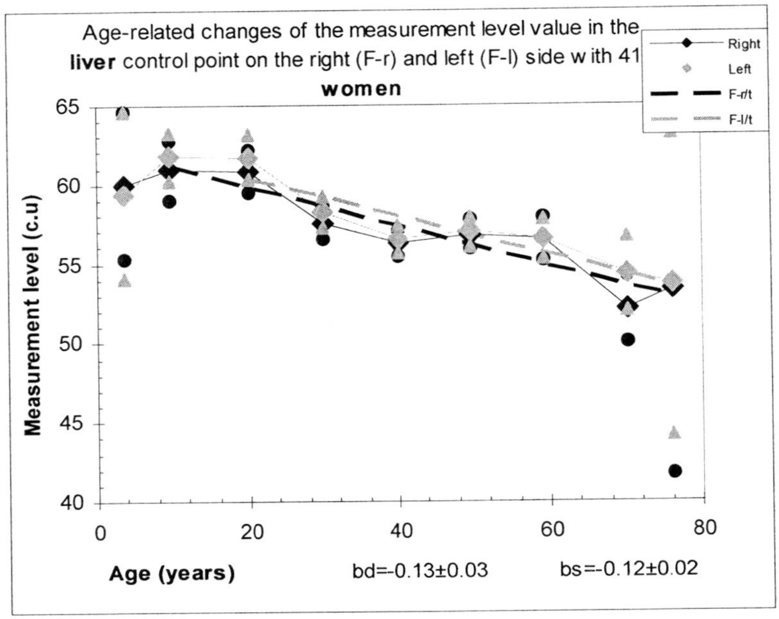

Picture 25.

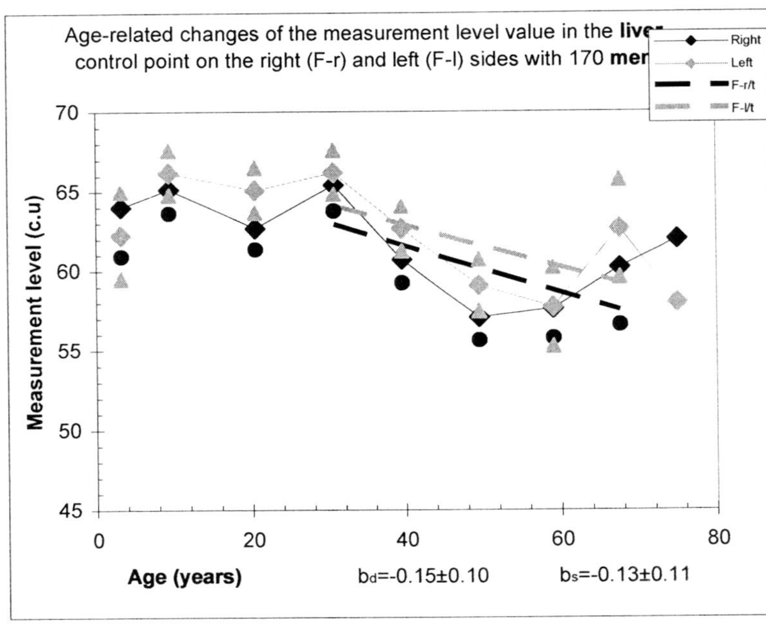

Picture 26.

Similar picture is observed with men (Picture 28.). Under the age of 30,36 a reliable increase from 57,25±1,52 c.u. for the right side and from 55,25±2,54 c.u. for the left side to 70,11±1,39 c.u. for the right side and to 70,74±1,10 c.u. for te left side is observed, then there is a plateau on the level of 61,13±4,07 c.u. for both sides within the age range of 39,56÷59,05 years, after that there is again a slowdown to 44 c.u. for the right side and 50 c.u. for the left side at the age of 75. Within the age range of 9,38÷30,63 years ML values remain above the physiological standard corridor. The inclinations of regression line equal to –0,33±0.07 for the right side and –0,29±0.07 for the left side.

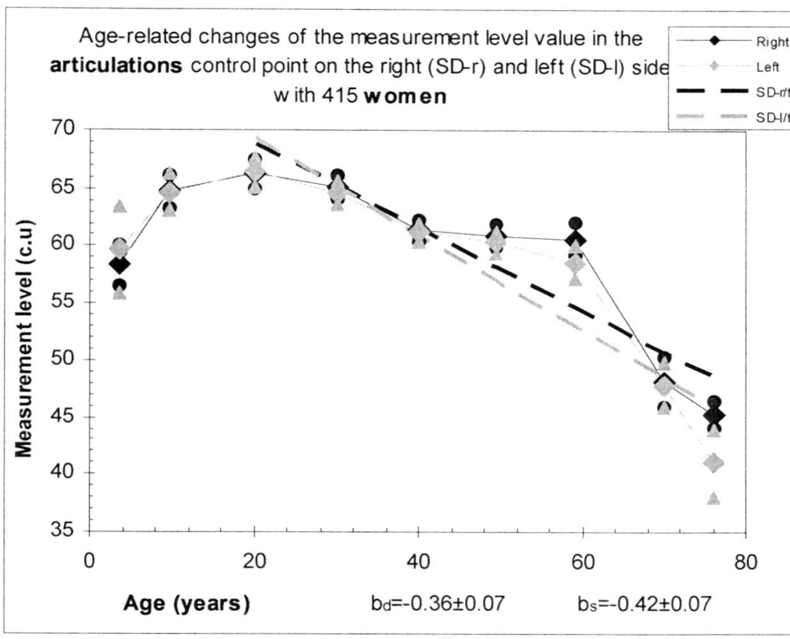

Picture 27.

2.1.15. Stomach

The stomach control point carries the information about 147 structural elements of stomach and alimentary tract including pyloric part, left part of stomach body, solar plexus, stomach peritoneum, aperture to pylorus and stomach body, fornix of ventricle, longitudal folds of stomach mucous coat along the small lesser curvature, cardinal part of stomach, body of stomach on the right.

Age-related changes of stomach ML can be divided into several stages (see Pictures 29÷30). The first stage of development or formation with women lasts till the age of 20,09, and woth men – till the age of 20,45 on the right and till the age of 30,63 on the left. Then a reliable decrease to 41,33±1,82 c.u. for the right side and to 40,67±1,73 c.u. fro the left side is observed with women at the age of 76.

Men also have a decrease of ML values to 43,00 c.u. on the right and 45,00 c.u. on the left, but at the same time there is an unreliable maximum at the age of 59,05 equaling to 58,32±1,9 c.u.. The inclinations of stomach regression lines for women equal to -0.32±0.05 for both sides, and for men -0.22±0.05 for the right side and -0.23±0.10 for the left side.

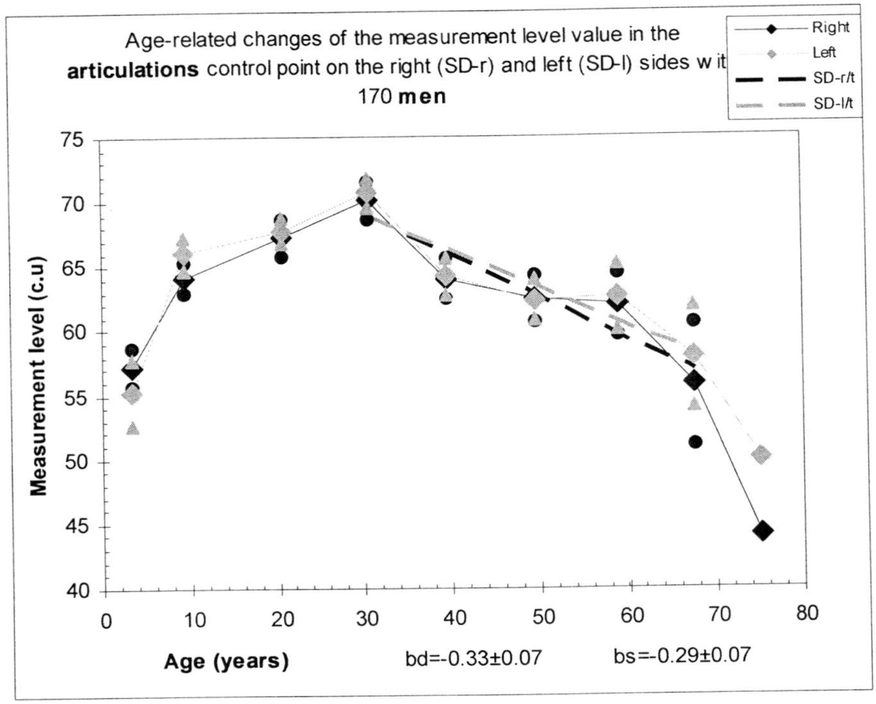

Picture 28.

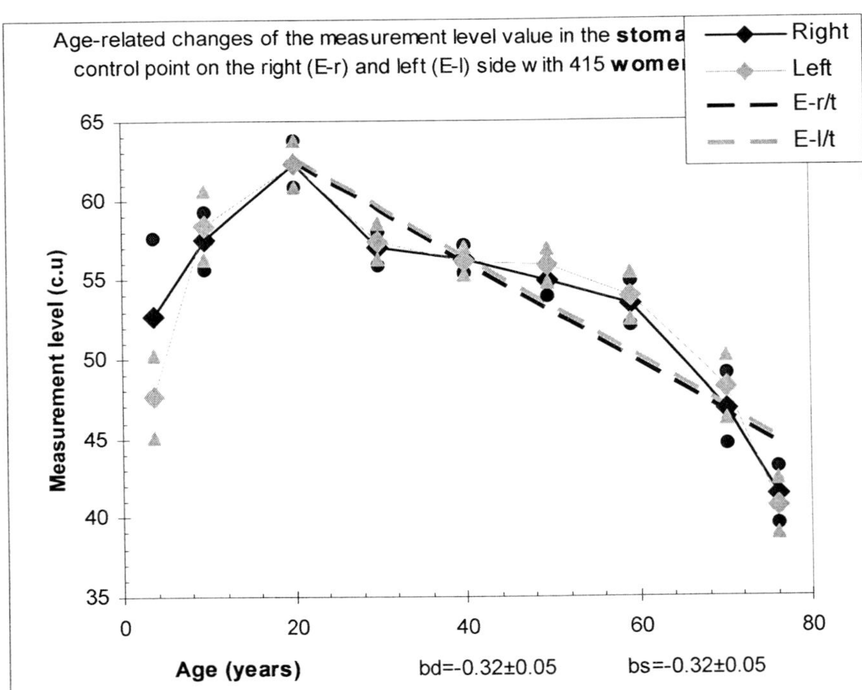

Picture 29.

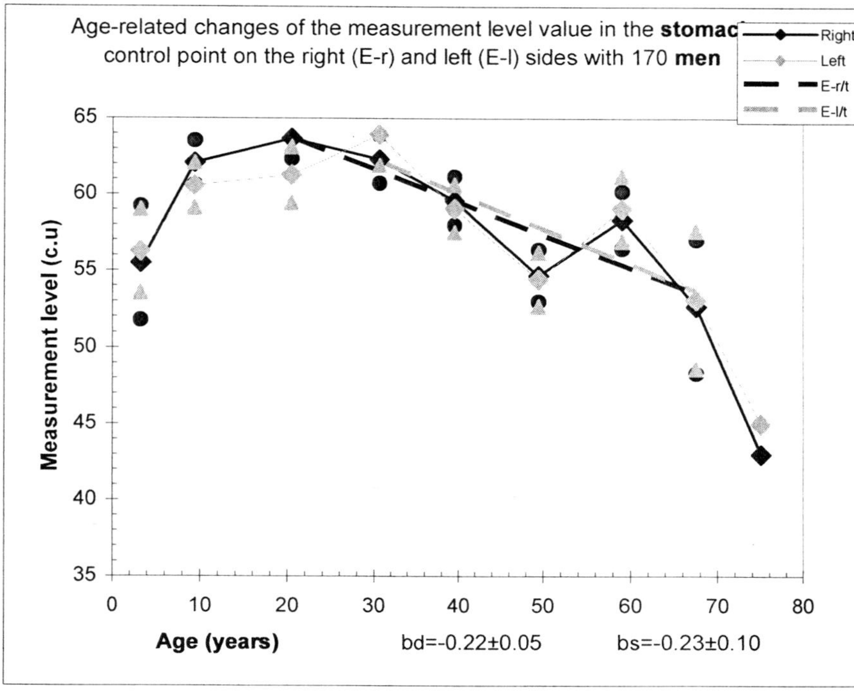

Picture 30.

2.1.16. Connective Tissue

The connective tissue control point carries generalized information about the following 5 parameters: connective tissue of abdominal cavity and small pelvis organs, connective tissue of body, mucous membrane degeneration, connective tissue of thorax and neck organs, connective tissue of head organs [2, 3].

With women an increase of measurement level from 53,33±2,79 c.u. for the right side and from 55,33±0,66 c.u. for the left side is observed at the age of 3,67 to 62,32±1,00 c.u. at the age of 9,63 for the right side and 60,74±0,99 c.u. at the age of 20,09 for the left side. Then there is a slowdown to 57,57±0,72 c.u. for the right side and to 57,43±0,62 c.u. by the age of 29,95 and a plateau on this level till the age of 59,19 for the right side and till the age of 49,53 for the left side. Then a slowdown of measurement level values to 45,33±4,17 c.u. is observed at the age of 76 (Picture 31.). Thereat, general decrease of measurement level by the age of 76 is reliable. The inclinations of regression lines equal to –0,20± 0,04 for the right side and –0,21±0,05 for the left side.

As for men, first, there is an increase of measurement level from 57,25±3,38 c.u. for the right side and from 55,50±1,17 c.u. for the left side at the age of 3,25 to 62,11±1,15 c.u. for the right side and to 62,84±0,90 c.u. for the left side at the age of 30,63. After that there is a decrease to 52,00 c.u. for both sides at the age of 75 (Picture 32.). However, only maximum index at the age of 30,63 for the left side is reliable. The inclinations of regression lines equal to –0,18± 0,02 for the right side and –0,21±0,02 for the left side.

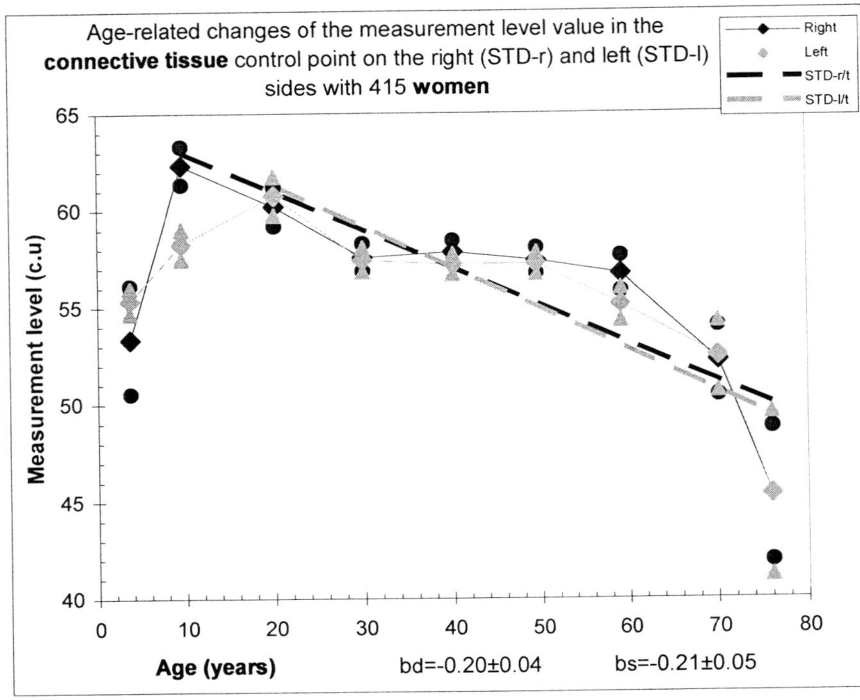

Picture 31.

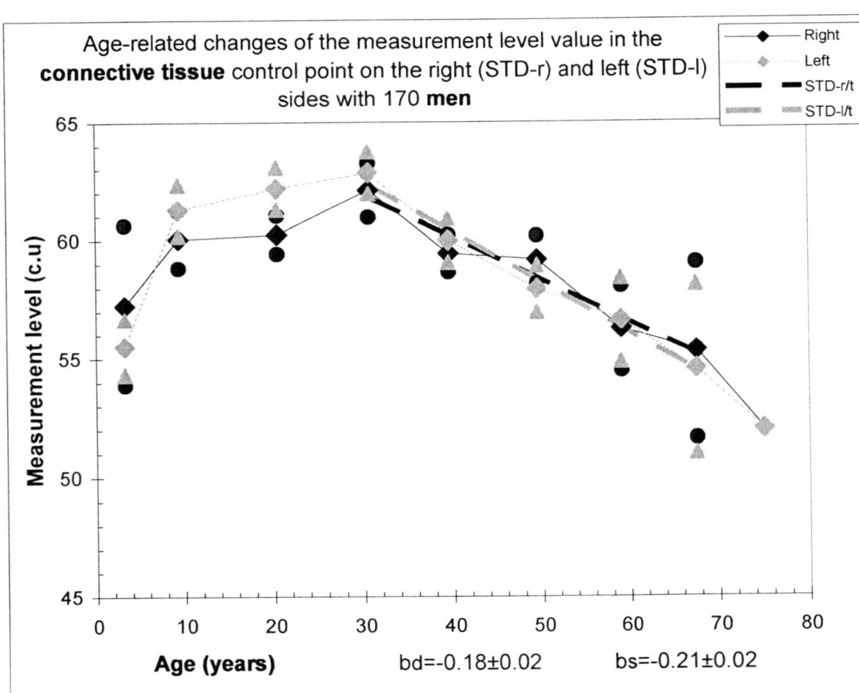

Picture 32.

2.1.17. Skin

The skin control point provides the information about 6 parameters, such as skin of lower body part and lower limbs, cicatricial skin changes, skin of upper body part including arm, nape and neck, head skin including hair follicles [1, 2].

As for women, there is an increase of measurement level observed till the age of 20,09, which is reliable only for the right side. At the age of 29,95 there is an unreliable minimum (55,71±1,01 c.u.) on the right, after which a plateau on the level of 54,03±3,60 c.u. for the right side and 56,02±1,97 c.u. for the left side is observed till the age of 59,19, and starting from the age of 70,10 there is a reliable decrease to 43,67±0,33 c.u. for the right side and 43,33±3,22 c.u. for the left side at the age of 76 (see Picture 33). The inclination of regression lines equals to –0,20±0,06 for both sides.

With men, there is the only maximum of all reliable differences discovered at the age of 30,63 for the right side (see Picture 34). The inclination of regression line for the right side – 0,17±0,04, and 0,07±0,02 for the left side.

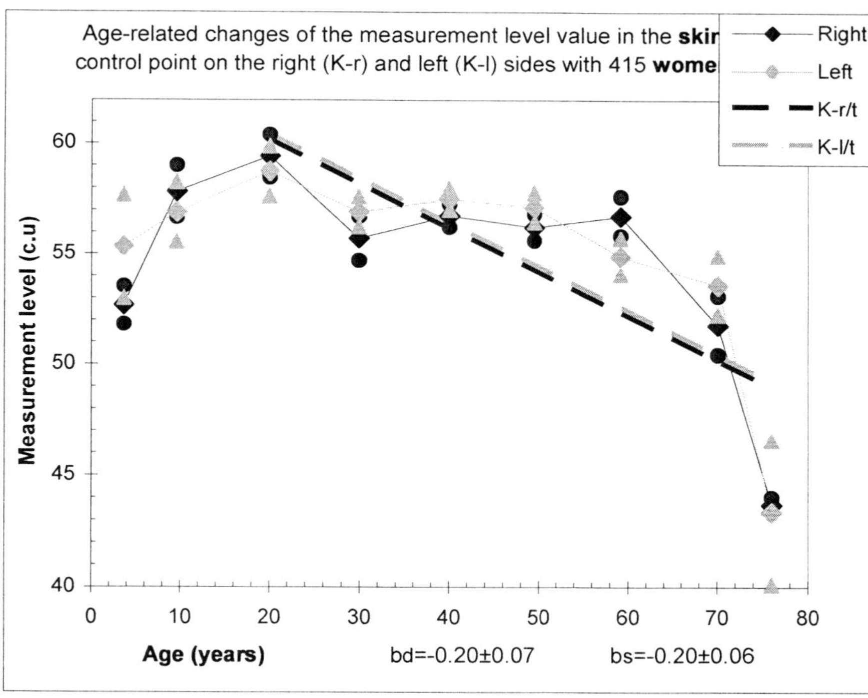

Picture 33.

2.1.18. Adipose Tissue

The adipose tissue control point carries the information about 4 parameters: adipose tissue of organs and abdominal cavity, adipose tissue of body, adipose tissue of thoracic cavity organs and vessels, head organs and vessels.

Female reliable age-related changes include an increase of measurement level till the age of 20.09 for the left side, ML drop after the age of 59,19 and maximum at the age of 20,09 for the right side and at the age of 29,95 for the left side (see Picture 35). The inclination of regression line on the right equals to –0,14±0,04, and on the left it is –0,20±0,07.

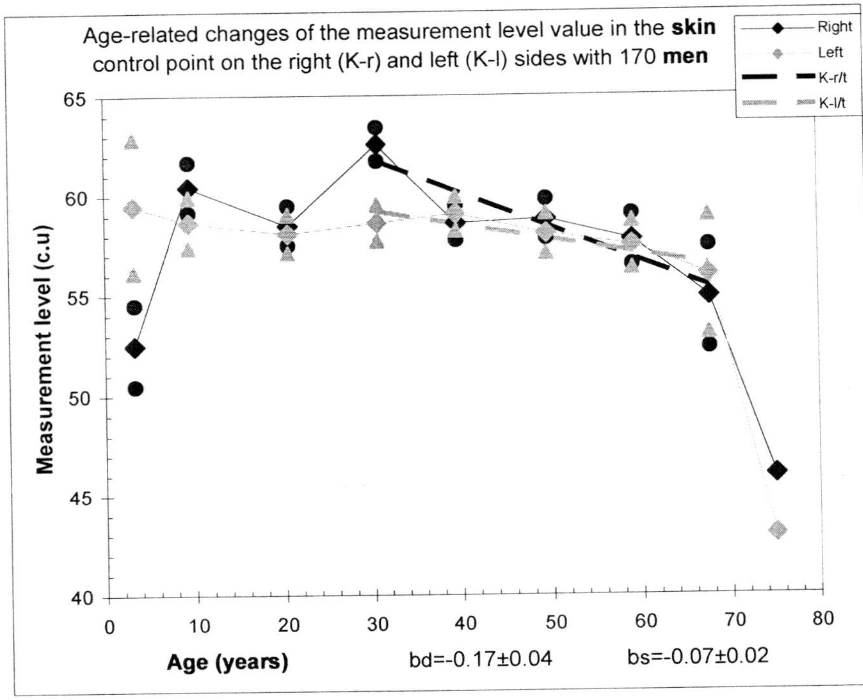

Picture 34.

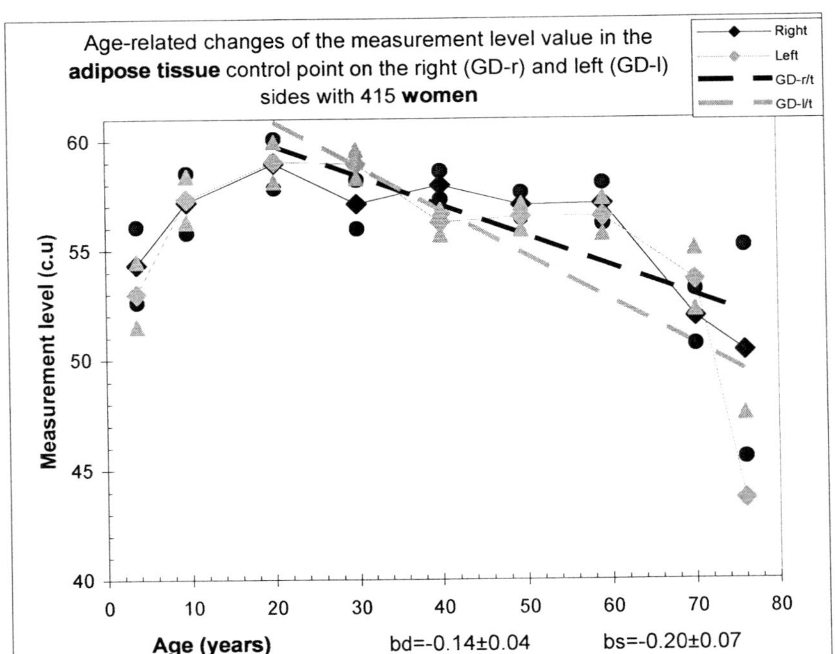

Picture 35.

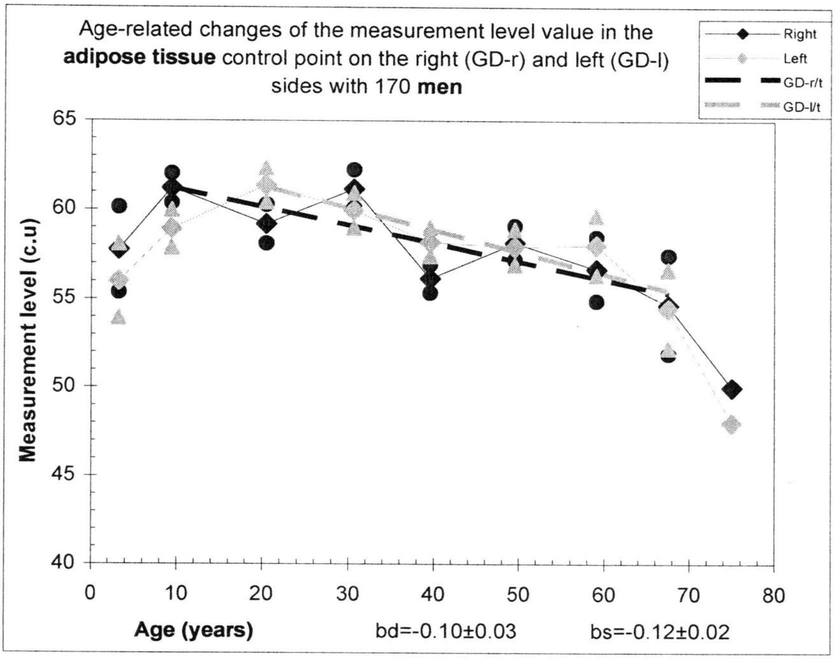

Picture 36.

As for men, there is a reliable difference for only one value of measurement level at the age of 20,45 on the left (see Picture 36). The inclination of regression line on the right equals to –0,10±0,03 and on the left it is –0,12±0,02.

2.1.19. Gallbladder

The gallbladder control point carries the information about 118 parameters, among them there are gall duct, common hepatic duct, gallbladder and bile passages, gallbladder peritoneum, cystic duct, left hepatic duct, gall ducts of right and left liver lobes [1, 2].

On Picture 37 for women one can see the below stages of age-related changes. Till the age of 20,09 there is an increase of measurement level value: from 49,33±1,64 c.u. to 56,49±1,54 y.e for the right side and from 50,00±3,96 c.u. to 55,57±1,65 c.u. for the left side, at that, the ML increase for the right side is reliable. Then a reliable decrease of measurement level is observed. The inclinations of regression lines equal to –0,20±0,03 for the right side and -0.17±0,03 for the left side.

As for men, only the maximum at the age of 9,38 for the right side, equaling to 59,58±1,38 c.u., is reliable of all differences in measurement level values. All other values range between 50÷56 units without reliable differences (Picture 38). The inclination of regression line on the right equals to –0,09±0,05, and on the left it is –0,10±0,07.

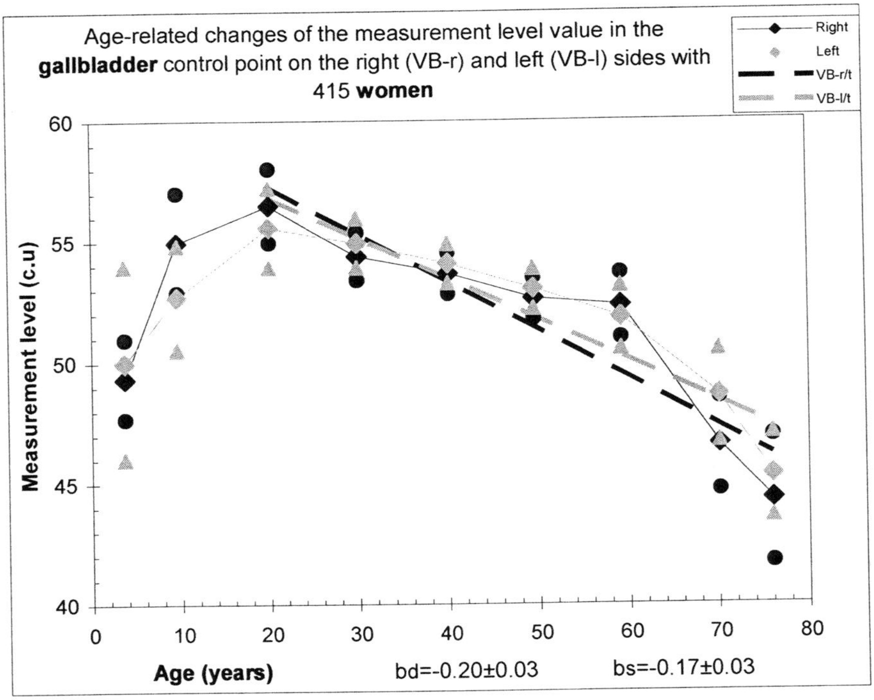

Picture 37.

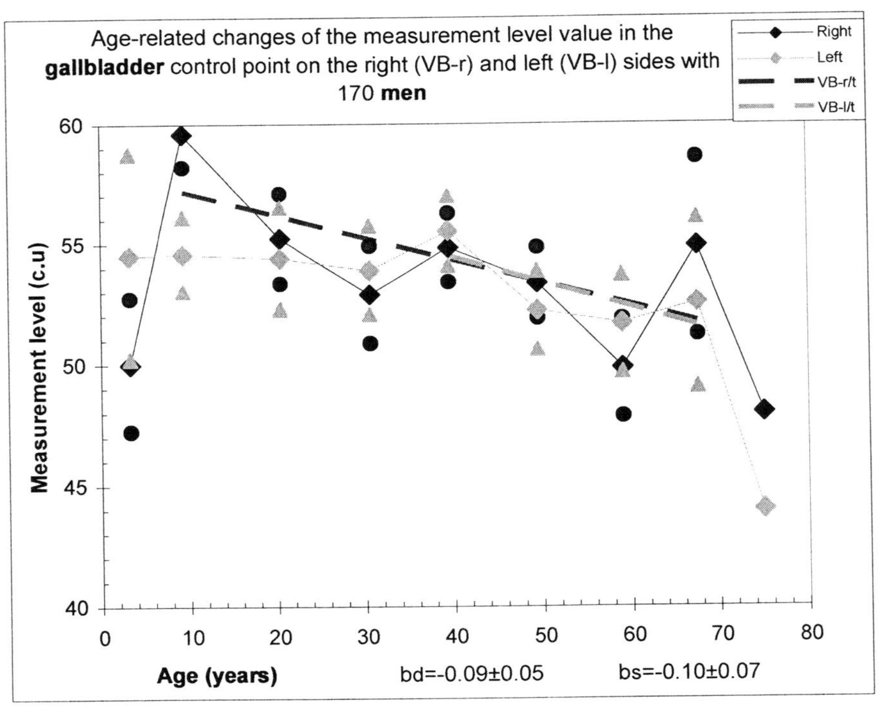

Picture 38.

2.1.20. Kidneys

The kidneys control point carries generalized information about 82 parameters related to kidneys and urinary system. Among the parameters there are renal pelvises, pyelorenalis zone, medullary substance of kidneys, kidney glomerules and tubules [1, 2].

With women from the age of 9,63 to 59,19 the values of measurement level range within the interval of 54÷58 c.u. for the right side and 55÷58 c.u. for the left side, at the same time a tendency to decrease is noticeable. After the age of 59,19 there is a reliable decrease of measurement level values to 42,00±2,99 c.u. for the right side and to 42,33±3,64 c.u. for the left side at the age of 76. Also, one can highlight the measurement level value for the left side at the age of 3,67 equal to 48,67±2,68 c.u. of reliable differences (Picture 39). The inclination of regression line for both sides equals to –0,18±0,06.

As for men, fluctuations of measurement level till the age of 39,56 are observed on the right and left within the interval of 53÷60 c.u., at that reliable maximums are discovered for both sides at the ages of 9,38 and 30,63. Then there is an unreliable decrease of measurement level values (see Picture 40). The inclination of regression line on the right equals to –0,16±0,07, and on the left it is –0,21±0,02.

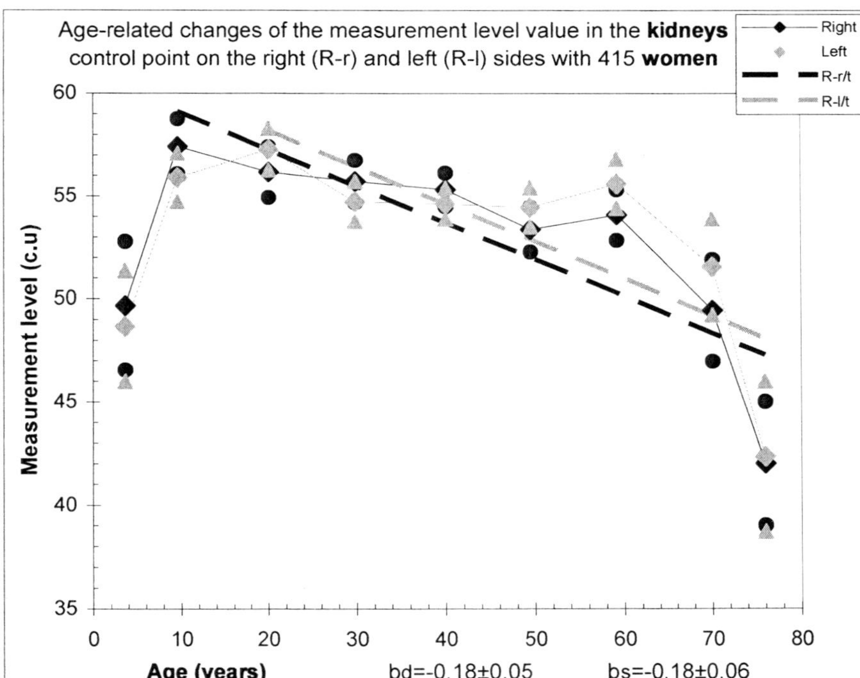

Picture 39.

2.1.21. Urogenital System

The urogenital system control point carries the information about 184 related parameters including such as body of urinary bladder, urinary bladder and urogenital organs, urinary bladder peritoneum, bottom and sphincter muscle of urinary bladder, prostate, penis, epididymis, uterine tube, lumbar spine and sacrum.

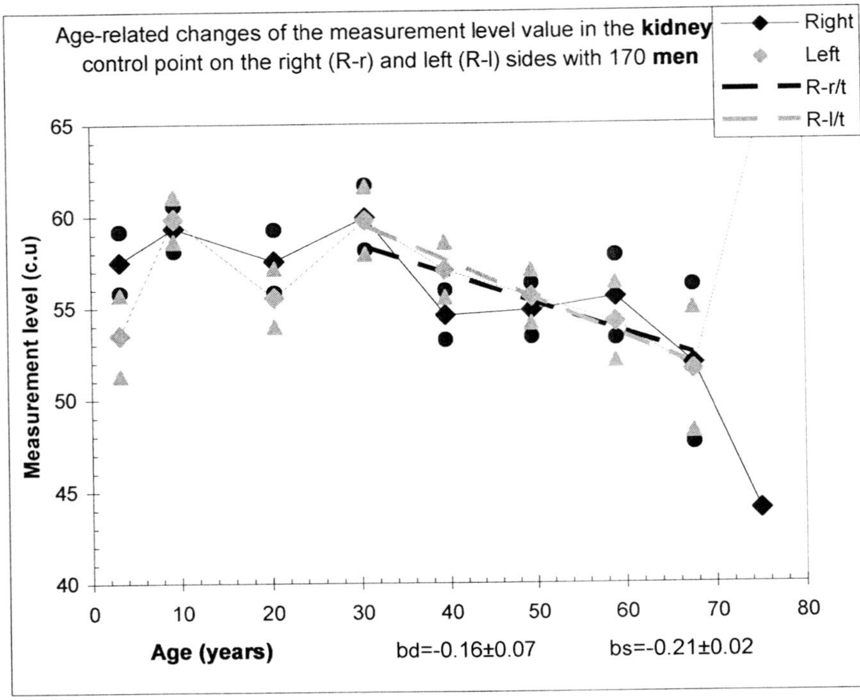

Picture 40.

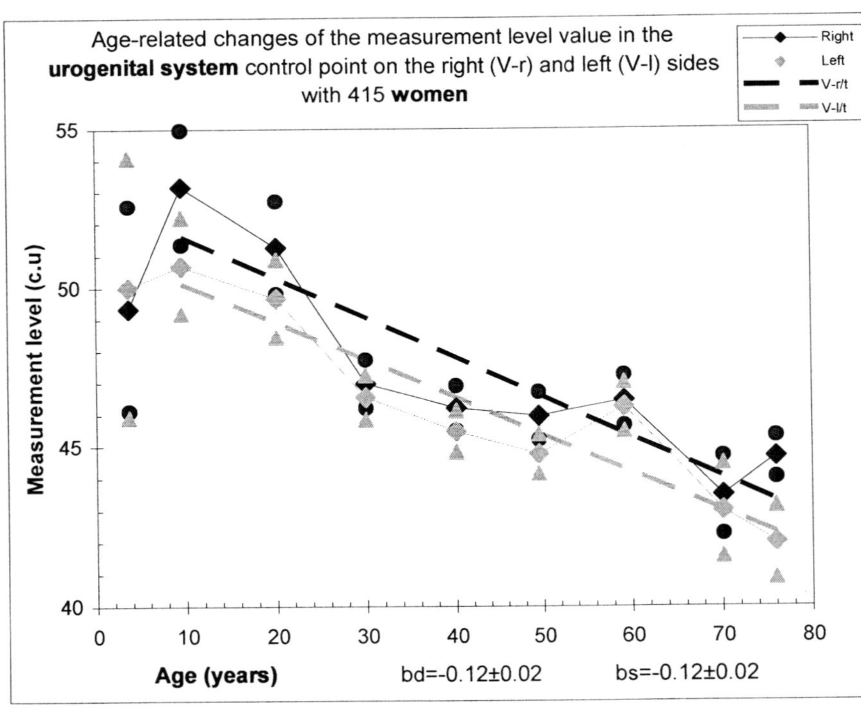

Picture 41.

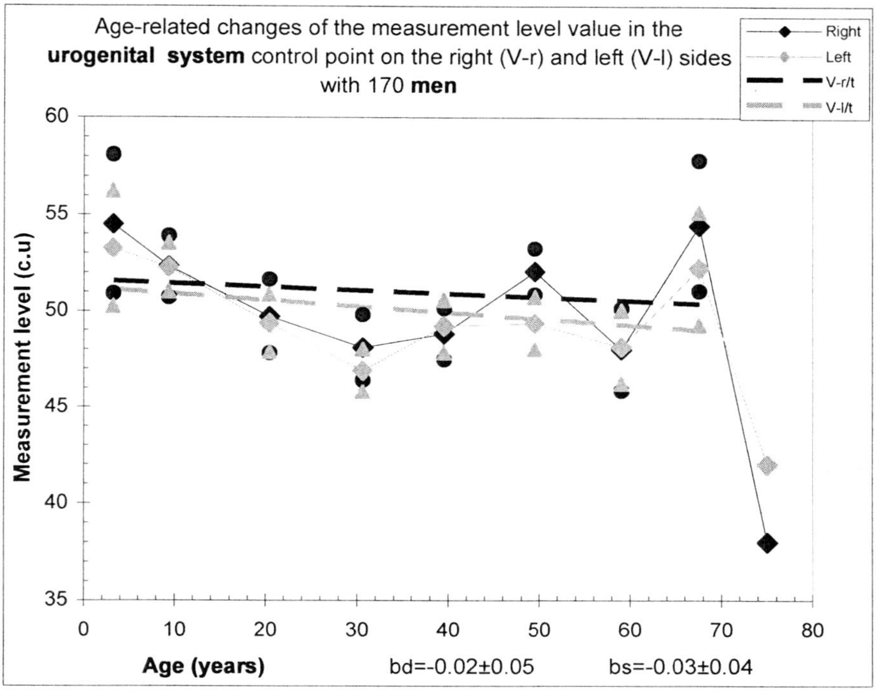

Picture 42.

For women the inclinations of regression lines for both sides equal to -0,12±0,02. Starting from the gae of 9,63 a decrease of measurement level value is observed. Maximum values of the measurement level are reliable for the two sides at the age of 9,63 and 20,09, and there is a reliable slowdown of this value at the age of 76 for the left side (see Picture 41). All measurement level values remain below the physiological standard corridor starting from the age of 29,9 for the right side andfrom the age of 20,09 for the left side.

Men have disclosed no reliable differences but, as it is seen from figure on Picture 42, one can suppose that the measurement level values stay at minimum at the age of 30.

Below there are tables of statistic values reflecting age-related changes for the examined organism systems with women (Table 4) and men (Table 5).

In these tables systems and organs are ranked in conformity with the values of regression coefficient for the right side of the body (descending). In column «*CA0*» there is an age index for each organism system, the index conforming to the initial point of regression line and plotted on the abscissa axis. In columns «*Ed0*» and «*Es0*» there are statistical (theoretical) measurement level values also correspond to initial points of regression line and are plotted on the ordinate axis. In columns «*bd*» and «*bs*» there are the values of regression coefficients, for the right and left sides of the body correspondingly, and in columns «td» and «ts» there are reliability values for these coefficients.

Table 4. Statistical values reflecting age-related changes for various systems of female organism

№	Systems of organism	Right side				Left side			
		CA0	*Ed0*	*bd*	*td*	*CA0*	*Es0*	*bs*	*ts*
1	Articulations	20,09	68,86	-0,36±0.07	5,01	20,09	69,28	-0,42±0,07	5,98
2	Stomach	20,09	62,45	-0,32±0.05	6,36	20,09	62,71	-0,32±0,05	5,94
3	Vascular system	9,63	53,77	-0,21±0.02	11,25	9,63	54,82	-0,21±0,02	13,25
4	Gallbladder	20,09	57,27	-0,20±0,03	5,79	20,09	56,82	-0,17±0,03	5,94
5	Immune system-3 (RPs)	---	---	---	---	9,63	60,73	-0,20±0,04	5,68
6	Connective tissue	9,63	62,98	-0,20±0,04	4,57	20,09	61,30	-0,21±0,05	4,71
7	Skin	20,09	60,17	-0,20±0,07	2,9	20,09	60,43	0,20±0,06±	3,18
8	Kidneys	9,63	59,13	-0,18±0,05	3,80	20,09	58,22	-0,18±0,06	2,84
9	Immune system-1 (AL)	20,09	55,00	-0,16±0,02	6,24	3,67	57,60	-0,18±0,03	7,13
10	Large intestine	9,63	49,58	-0,14±0,01	12,37	9,63	49,84	-0,13±0,01	9,07
11	Immune system-2 (PAD)	9,63	58,20	-0,14±0,01	10,90	3,67	58,04	-0,12±0,01	13,42
12	Heart	20,09	59,76	-0,14±0,03	5,46	20,09	58,29	-0,13±0,01	15,06
13	Adipose tissue	20,09	59,76	-0,14±0,04	3,84	20,09	60,87	-0,20±0,07	3,84
14	Endocrine system	20,09	51,85	-0,13±0,02	6,19	20,09	53,44	-0,18±0,04	5,12
15	Liver	9,63	61,19	-0,13±0,03	4,83	9,63	60,39	-0,12±0,02	5,77
16	Urogenital system	9,63	51,60	-0,12±0,02	5,14	9,63	50,13	-0,12±0,02	6,49
17	Nervous system	20,09	49,44	-0,12±0,03	3,98	9,63	50,20	-0,13±0,02	5,34
18	Small intestine	9,63	56,11	-0,11±0,04	2,99	9,63	58,41	-0,13±0,03	4,03
19	Respiratory system	20,09	58,60	-0,10±0,04	2,43	20,09	58,50	-0,11±0,03	3,73
20	Pancreas	20,09	63,84	-0,10±0,04	2,19	---	---	---	---
21	Lymphoid system	20,09	59,14	-0,06±0,03	2,12	20,09	60,87	-0,09±0,02	4,35

Table 5. Statistical values reflecting age-related changes for various systems of male organism

№	Systems of organism	Right side				Left side			
		CA0	*Ed0*	*bd*	*td*	*CA0*	*Es0*	*bs*	*ts*
1	Articulations	30,63	68,94	-0,33±0,07	4,76	30,63	68,99	-0,29±0,07	4,16
2	Immune system-1 (AL)	30,63	58,30	-0,26±0,05	5,20	30,63	58,33	-0,26±0,06	4,49
3	Heart	30,63	62,74	-0,25±0,04	6,57	30,63	60,31	-0,25±0,02	11,21
4	Stomach	20,45	63,74	-0,22±0,05	4,09	30,63	62,22	-0,23±0,10	2,24
5	Small intestine	30,63	58,36	-0,20±0,04	4,49	30,63	60,22	-0,19±0,05	3,62
6	Vascular system	3,25	58,86	-0,19±0,03	5,43	9,38	58,12	-0,22±0,03	7,26
7	Connective tissue	30,63	61,79	-0,18±0,02	7,57	30,63	62,36	-0,21±0,02	12,67
8	Pancreas	30,63	68,54	-0,18±0,07	2,7	----	---	----	----
9	Immune system-3 (RPs)	---	---	---		30,63	63,22	-0,16±0,06	2,63
10	Large intestine	9,38	53,59	-0,17±0,02	8,48	9,38	53,85	-0,16±0,03	5,32
11	Skin	30,63	61,75	-0,17±0,04	4,20	3,25	59,23	-0,07±0,02	4,38
12	Nervous system	3,25	54,99	-0,16±0,02	6,59	9,38	53,39	-0,14±0,03	4,17
13	Kidneys	30,63	58,31	-0,16±0,07	2,39	9,38	59,44	-0,21±0,02	11,43
14	Immune system-2 (PAD)	3,25	60,94	-0,15±0,03	5,22	30,63	58,69	-0,22±0,04	5,68
15	Endocrine system	3,25	58,78	-0,15±0,03	5,20	3,25	60,67	-0,19±0,04	4,96
16	Liver	30,63	62,95	-0,15±0,10	1,5	30,63	64,13	-0,13±0,11	1,24
17	Adipose tissue	9,38	61,21	-0,10±0,03	3,50	20,45	61,25	-0,12±0,02	4,93
18	Respiratory system	30,63	60,04	-0,10±0,09	1,20	30,63	61,79	-0,15±0,08	1,81
19	Gallbladder	9,38	57,18	-0,09±0,05	1,97	39,56	54,50	-0,10±0,07	1,52
20	Lymphoid system	30,63	61,80	-0,08±0,07	1,09	30,63	63,20	-0,14±0,03	4,84
21	Urogenital system	3,25	51,57	-0,02±0,05	0,39	3,25	51,13	-0,03±0,04	0,89

2.2. Overall Picture of Ageing

We have studied general age-related changes with men and women. On Pictures 43, 44 there are figures reflecting those changes. Figure analysis for men and women allowed to single out 2 stages reflecting ontogenesis of vitally important organism systems: development stage and regression or ageing stage.

On the figures it is evident that with women age-related regression process starts on average from the age of 20,09. But for men the period of development and optimal physiological activity lasts longer, and the regression process starts at the age of 30,63.

Below there are reliability tables for general age-related changes according to t-criterium for men and women (see Table 6÷7). Each table has diagonal separation. In the right corner part of the table difference reliability values of measurement levels of age groups for the right side of the body are shown, and in the left corner part there are difference reliability values of measurement levels for the left side. Two top lines and two left columns of the table enumerate the examined age groups (Δx) and average age for each of them ($\bar{x}$). The values of reliable age-related changes are marked with bold font provided that the significance level is P<0,05. From the tables one can figure out that with women practically all measurement level values differ reliably, and with men – most of them differ.

Below there is a table of statistic values reflecting general age-related changes for men and women (see Table 8).

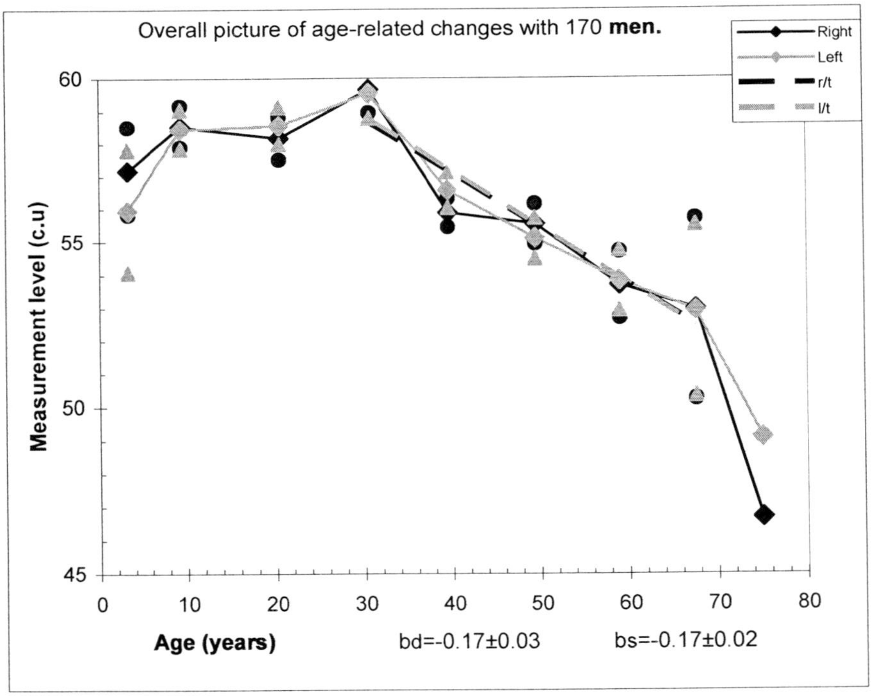

Picture 43.

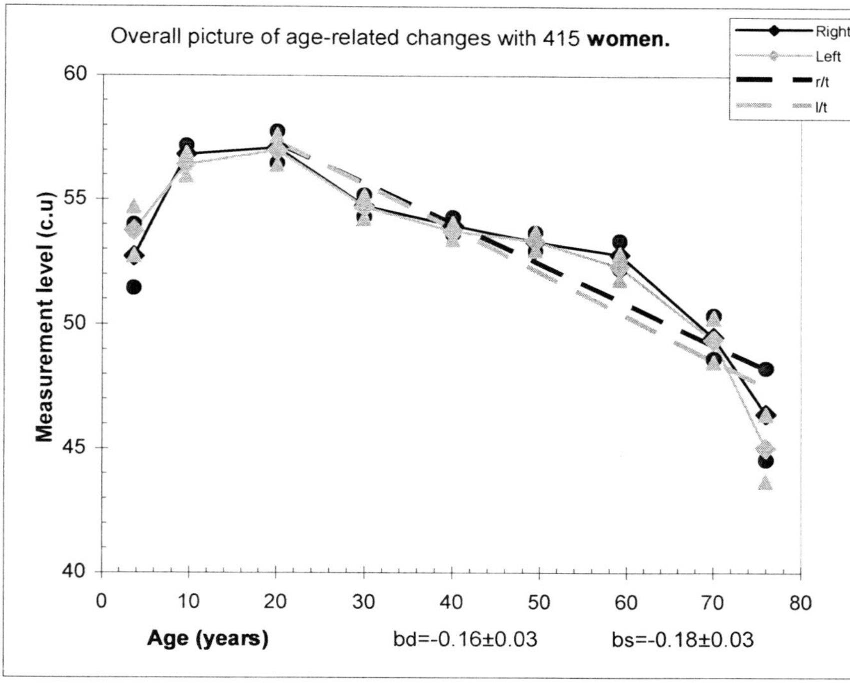

Picture 44.

Table 6. Reliability of general age-related changes of right and left sides of male body

R\L	Δx	0-4	5-14	15-24	25-34	35-44	45-54	55-64	65-74	75-84
Δx	$\bar{x}$	3,25	9,38	20,45	30,63	39,56	49,51	59,05	67,56	75,00
0-4	3,25		0,89	0,66	1,61	0,91	1,09	2,03	1,35	
5-14	9,38	1,25		0,66	1,18	3,38	3,33	3,99	1,94	
15-24	20,45	1,31	0,130		1,51	2,86	2,87	3,66	1,81	
25-34	30,63	1,75	1,11	1,02		4,52	4,36	4,79	2,32	
35-44	39,56	0,32	2,27	2,47	3,13		0,87	1,95	1,03	
45-54	49,51	0,41	3,79	4,00	4,45	1,74		1,54	0,90	
55-64	59,05	0,99	4,15	4,31	4,74	2,54	1,14		0,25	
65-74	67,56	0,93	2,03	2,08	2,40	1,35	0,81	0,33		
75-84	75,00									

Table 7. Reliability of general age-related changes of right and left sides of female body

R\L	Δx	0-4	5-14	15-24	25-34	35-44	45-54	55-64	65-74	75-84
Δx	$\bar{x}$	3,67	9,63	20,09	29,95	40,07	49,53	59,19	70,10	76,00
0-4	3,67		3,05	3,01	1,48	0,93	0,44	0,06	2,05	2,75
5-14	9,63	2,44		3,01	3,59	5,84	6,76	6,03	7,70	5,41
15-24	20,09	2,78	0,75		2,96	4,28	5,02	4,98	6,93	5,34
25-34	29,95	0,89	2,60	2,95		1,42	2,46	2,71	5,32	4,29
35-44	40,07	0,00	4,74	4,68	1,72		1,29	1,79	4,75	3,93
45-54	49,53	0,37	5,17	5,07	2,29	0,79		0,78	4,01	3,58
55-64	59,19	1,26	5,88	5,79	3,42	2,30	1,60		3,17	3,24
65-74	70,10	3,20	6,87	6,91	5,19	4,48	4,01	2,77		1,48
75-84	76,00	5,08	7,78	7,87	6,60	6,10	5,78	4,90	2,62	

Table 8. Statistical values reflecting general age-related changes for men and women

Sex	Right side				Left side			
	CA0	*Ed0*	*bd*	*td*	*CA0*	*Es0*	*bs*	*ts*
Female	20,09	57,27	-0,16±0,03	6,31	20,09	57,36	-0,18±0,03	5,63
Male	30,63	58,63	-0,17±0,03	4,96	30,63	58,65	-0,17±0,03	7,26

2.3. Biological Age Determination

On the basis of rearrangement of rectilinear regression equation we have derived a formula to determine biological age in a general view:

$$BA = CA_0 + \frac{\frac{1}{-b_d}(E_{d0} - E_d) + \frac{1}{-b_s}(E_{s0} - E_s)}{2} \quad (1)$$

where *BA* – is the value of biological age of a certain person,

CA_0 – is a statistic age plotted on abscissa axis and conforming to the initial point of theoretic regression line;

E_{d0} and E_{s0} - statistic values of electrical conductivity on the right and left sides of the body, the values being plotted on ordinate axis and conforming to the initial point of theoretic regression line;

E_d and E_s - measured electrical conductivity values on the right and left side of the body of a certain person,

b_d and b_s - regression coefficient of age-related changes on the right and left sides of the body

Placing regression constants (statistic values) from Table 8 in formula (1), we have derived formulas to determine BA for men and for women.

$$\text{for men } BA = 30{,}63 + \frac{5{,}88(58{,}65 - E_{d1}) + 5{,}88(58{,}65 - E_{s1})}{2}, \qquad (2)$$

$$\text{for women: } BA = 20{,}09 + \frac{6{,}25(57{,}27 - E_{d1}) + 5{,}55(57{,}36 - E_{s1})}{2}, \qquad (3)$$

where E_{d1} and E_{s1} - average values of the measured one in right and left control points.

Such qualitative evaluation of BA is a simple and quick variant of BA calculation, but for all that separate systems and organs are not distinguished, which reduces the information value of the method significantly.

The second variant of BA calculation has more valuable in terms of information, as it allows to determine BA or ageing speed as well as the degree of pathological deflections for each system and organ being examined.

Below there are formulas to determine BA for various systems of the organism taking into account right and left sides of the body for men (see formulas 4÷23) and women (see formulas 24÷44):

for men:

for lymphoid system* (BA_{li}): BA_{li} = 30,63+7,14(63,20- LI_s), (4)

for respiratory system* (BA_p): BA_p = 30,63+6,67(61,79- P_s), (5)

for large intestine (BA_{gi}): (6)

$$BA_{gi} = 9{,}38 + \frac{5{,}88(53{,}59 - GI_d) + 6{,}25(53{,}85 - GI_s)}{2},$$

for nervous system (7)

$$(BA_{nd}): BA_{nd} = 6{,}32 + \frac{6{,}25(54{,}99 - ND_d) + 7{,}14(53{,}39 - ND_s)}{2}$$

for vascular system ($BA_{мс}$): $BA_{мс}$ = (8)

$$6{,}32+\frac{5{,}26(58{,}86-MC_d)+4{,}55(58{,}12-MC_s)}{2},$$

for immune system-1 (9)

(BA_{al}): $$BA_{al}=30{,}63+\frac{3{,}85(58{,}30-AL_d)+3{,}85(58{,}33-AL_s)}{2},$$

for immune system -2 (10)

(BA_{pad}): $$BA_{pad}=16{,}94+\frac{6{,}67(60{,}94-PAD_d)+4{,}55(58{,}69-PAD_s)}{2},$$

for endocrine system (BA_{tr}): BA_{tr} = (11)

$$3{,}25+\frac{6{,}67(58{,}78-TR_d)+5{,}26(60{,}67-TR_s)}{2},$$

for heart (BA_c): $$BA_c = 30{,}63+\frac{4{,}0(62{,}74-C_d)+4{,}0(60{,}31-C_s)}{2},\quad(12)$$

for small intestine (BA_{ig}): BA_{ig} = (13)

$$30{,}63+\frac{5{,}0(58{,}36-IG_d)+5{,}26(60{,}22-IG_s)}{2},$$

for pancreas (BA_{rpd}): $BA_{rpd} = 30{,}63+5{,}56(68{,}54- RP_d)$, (14)

for immune system-3 or spleen (BA_{rps}): $BA_{rps} = 30{,}63+6{,}25(63{,}22- RP_s)$, (15)

for liver (BA_f): $$BA_f = 30{,}63+\frac{6{,}67(62{,}95-F_d)+7{,}69(64{,}13-F_s)}{2},\quad(16)$$

for articulations (17)

(BA_{sd}): $$BA_{sd}=30{,}63+\frac{3{,}03(68{,}94-SD_d)+3{,}45(68{,}99-SD_s)}{2},$$

for stomach (BA_e): BA_e = (18)

$$25{,}54+\frac{4{,}55(63{,}74-E_d)+4{,}35(62{,}22-E_s)}{2},$$

for connective tissue (19)

$$(BA_{std}): BA_{std} = 30{,}63 + \frac{5{,}56(61{,}79 - STD_d) + 4{,}76(62{,}36 - STD_s)}{2},$$

for skin (BA_k): (20)

$$BA_k = 16{,}94 + \frac{5{,}88(61{,}75 - K_d) + 14{,}28(59{,}23 - K_s)}{2},$$

for adipose tissue (21)

$$(BA_{gd}): BA_{gd} = 14{,}92 + \frac{10{,}00(61{,}21 - GD_d) + 8{,}33(61{,}25 - GD_s)}{2},$$

for gallbladder (BA_{vb}): BA_{vb} (22)

$$= 24{,}47 + \frac{11{,}11(57{,}18 - VB_d) + 10{,}00(54{,}50 - VB_s)}{2},$$

for kidneys (23)

$$(BA_r): BA_r = 20{,}01 + \frac{6{,}25(58{,}31 - R_d) + 4{,}76(59{,}44 - R_s)}{2},$$

* - in the formulas of these systems, BA is calculated only for the left side of the body because age-related changes of the measured values for the right side are unreliable.

for women:

for lymphoid system (24)

$$(BA_{li}): BA_{li} = 20{,}09 + \frac{16{,}66(59{,}14 - LI_d) + 11{,}11(60{,}87 - LI_s)}{2},$$

for respiratory system (BA_p): $BA_p =$ (25)

$$20{,}09 + \frac{10{,}0(58{,}6 - P_d) + 9{,}09(58{,}50 - P_s)}{2},$$

for large intestine (BA_{gi}): (26)

$$BA_{gi} = 9{,}63 + \frac{7{,}14(49{,}58 - GI_d) + 7{,}69(49{,}84 - GI_s)}{2},$$

for nervous system (27)

$$(BA_{nd}): BA_{nd} = 14{,}86 + \frac{8{,}33(49{,}44 - ND_d) + 7{,}69(50{,}22 - ND_s)}{2},$$

for vascular system ($BA_{мс}$): $BA_{мс}$ = (28)

$$9{,}63+\frac{4{,}76(53{,}77-MC_d)+4{,}76(54{,}82-MC_s)}{2},$$

for immune system-1 (29)

(BA_{al}): BA_{al} = $11{,}88+\frac{6{,}25(55{,}0-AL_d)+5{,}56(57{,}60-AL_s)}{2}$,

for immune system -2 (30)

(BA_{pad}): BA_{pad} = $6{,}65+\frac{7{,}14(58{,}2-PAD_d)+8{,}33(58{,}04-PAD_s)}{2}$,

for endocrine system (BA_{tr}): BA_{tr} = (31)

$$20{,}09+\frac{7{,}69(51{,}85-TR_d)+5{,}55(53{,}44-TR_s)}{2},$$

for heart (BA_c): BA_c = $20{,}09+\frac{7{,}14(59{,}76-C_d)+7{,}69(58{,}29-C_s)}{2}$, (32)

for small intestine (BA_{ig}): BA_{ig} = (33)

$$9{,}63+\frac{9{,}09(56{,}11-IG_d)+7{,}69(58{,}41-IG_s)}{2},$$

for pancreas (BA_{rpd}): BA_{rpd} = $20{,}09+10{,}0(63{,}84-RP_d)$, (34)

for immune system-3 or spleen (BA_{rps}): BA_{rps} = $9{,}63+5{,}0(60{,}73-RP_s)$, (35)

for liver (BA_f): BA_f = $9{,}63+\frac{7{,}69(61{,}19-F_d)+8{,}33(60{,}39-F_s)}{2}$, (36)

for articulations

(BA_{sd}): BA_{sd} = $20{,}09+\frac{2{,}77(68{,}86-SD_d)+2{,}38(69{,}28-SD_s)}{2}$,

(37)

for stomach (BA_e): BA_e =

(38)

$$20{,}09+\frac{3{,}12(62{,}45-E_d)+3{,}12(62{,}71-E_s)}{2},$$

for connective tissue

$$(BA_{std}): BA_{std} = 14{,}86 + \frac{5{,}0(62{,}98 - STD_d) + 4{,}76(61{,}3 - STD_s)}{2}, \quad (39)$$

$$\text{for skin } (BA_k): BA_k = 20{,}09 + \frac{5{,}0(60{,}17 - K_d) + 5{,}0(60{,}43 - K_s)}{2}, \quad (40)$$

for adipose tissue

$$(BA_{gd}): BA_{gd} = 20{,}09 + \frac{7{,}14(59{,}76 - GD_d) + 5{,}0(60{,}87 - GD_s)}{2}, \quad (41)$$

for gallbladder (BA_{vb}): $BA_{vb} =$

$$20{,}09 + \frac{5{,}0(57{,}27 - VB_d) + 5{,}88(56{,}82 - VB_s)}{2}, \quad (42)$$

for kidneys

$$(BA_r): BA_r = 14{,}86 + \frac{5{,}56(59{,}13 - R_d) + 5{,}56(58{,}22 - R_s)}{2}, \quad (43)$$

for urogenital system

$$(BA_v): BA_v = 9{,}63 + \frac{8{,}33(51{,}60 - V_d) + 8{,}33(50{,}13 - V_s)}{2}, \quad (44)$$

where:

LI_d and LI_s – measured values of electrical conductivity on the right and left branches of lymphatic meridian;

P_d and P_s – measured values of electrical conductivity on the right and left branches of lungs meridian;

GI_d and GI_s – measured values of electrical conductivity on the right and left branches of large intestine meridian;

ND_d and ND_s – measured values of electrical conductivity on the right and left branches of nervous degeneration meridian;

MC_d and MC_s – measured values of electrical conductivity on the right and left branches of pericardium meridian;

AL_d and AL_s – measured values of electrical conductivity on the right and left branches of allergy meridian;

PAD_d and PAD_s – measured values of electrical conductivity on the right and left branches of epithelium and parenchymatous degeneration meridian;

TR_d and TR_s – measured values of electrical conductivity on the right and left branches of triple heater meridian;

C_d and C_s – measured values of electrical conductivity on the right and left branches of heart meridian;

IG_d and IG_s – measured values of electrical conductivity on the right and left branches of small intestine meridian;

RP_d – measured value of electrical conductivity on the right branch of pancreas-spleen meridian;

RP_s – measured value of electrical conductivity on the left branch of pancreas-spleen meridian;

F_d and F_s – measured values of electrical conductivity on the right and left branches of liver meridian.

SD_d and SD_s – measured values of electrical conductivity on the right and left branches of articular degeneration meridian;

E_d and E_s – measured values of electrical conductivity on the right and left branches of stomach meridian;

STD_d and STD_s – measured values of electrical conductivity on the right and left branches of connective tissue degeneration meridian;

K_d and K_s – measured values of electrical conductivity on the right and left branches of skin meridian;

GD_d and GD_s – measured values of electrical conductivity on the right and left branches of adipose degeneration meridian;

VB_d and VB_s – measured values of electrical conductivity on the right and left branches of gallbladder meridian;

R_d and R_s – measured values of electrical conductivity on the right and left branches of kidneys meridian;

V_d and V_s – measured values of electrical conductivity on the right and left branches of urinary bladder meridian.

One of the necessary conditions to increase the quality of BA estimation is independence from pathological processes accompanying ageing. The present method of BA determination provides this condition by means of determining the systems with pathological deflections and excluding them from further calculation.

The following variant to determine pathological deflections for separate systems of the organism is proposed:

$$S_{pat_i} = \frac{|BA_c - BA_i|}{\sigma\sqrt{\frac{n}{n-1}}}, \tag{45}$$

where S_{pat_i} – degree of pathological deflections;

BA_i – BA value for a separate system of the organism (BA_{li} , BA_p , BA_{gi} ,... BA_v);

BA_c – average of all BA values of separate systems (without exception);

σ - average quadratic deflection for a range of all BA values of separate systems;

n – number of all analyzed systems

If the value of pathological deflection degree is above or equal to 1,5 ($S_{pat_i} \geq 1{,}5$), it is assumed that the analyzed system has pathological deflections which can influence BA estimation corrupting its values.

Then we derive BA values average for separate systems excluding the systems with pathological deflections with men and women according to the formula:

$$BA = \frac{\sum BA_i}{n}, \tag{46}$$

where BA_i – is a BA value of a separate system (BA_{li} , BA_p , BA_{gi} ,... BA_v);

n – number of organism systems without pathological deflections.

2.4. Example of Method Usage (Computer Variant)

Name,	Surname	Patient № 252	test date	01-Sep-2005	CA=56

Test result	Your biological age is 54,15 years You are younger than your calendar age by 1,85 years Average speed of age-related changes is 96,70%

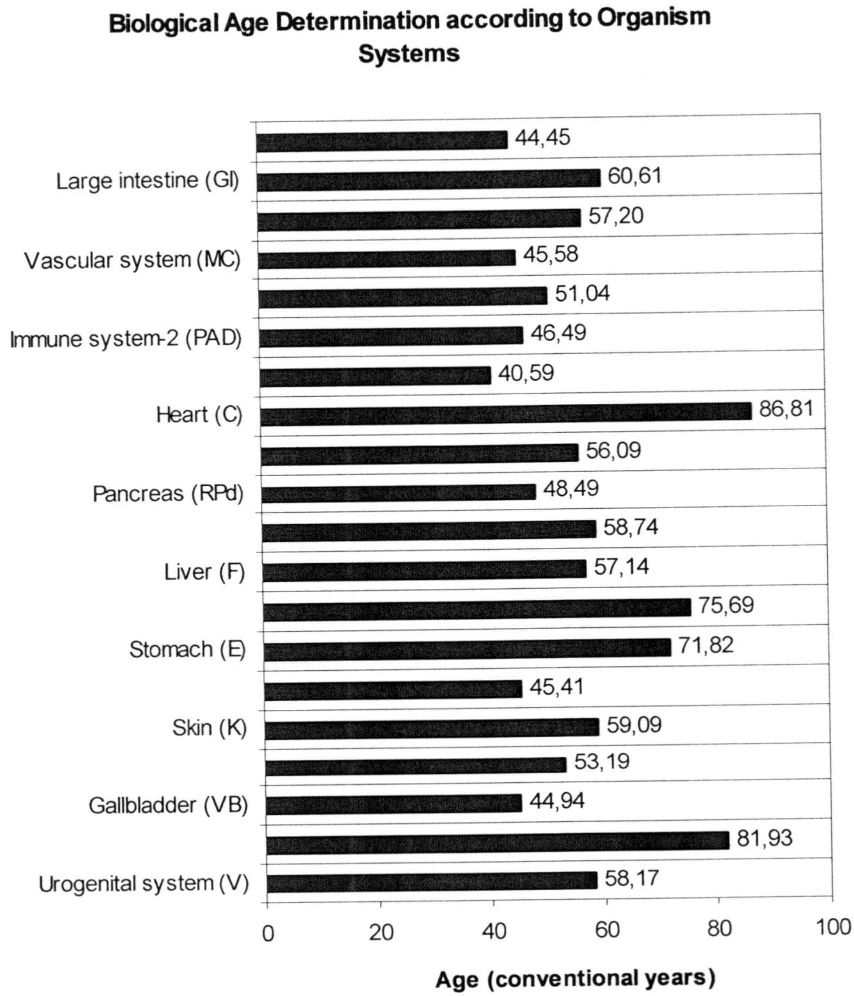
Biological Age Determination according to Organism Systems
44,45
Large intestine (GI)
60,61
57,20
Vascular system (MC)
45,58
51,04
Immune system-2 (PAD)
46,49
40,59
Heart (C)
86,81
56,09
Pancreas (RPd)
48,49
58,74
Liver (F)
57,14
75,69
Stomach (E)
71,82
45,41
Skin (K)
59,09
53,19
Gallbladder (VB)
44,94
81,93
Urogenital system (V)
58,17
0
20
40
60
80
100
Age (conventional years)

Picture 45.

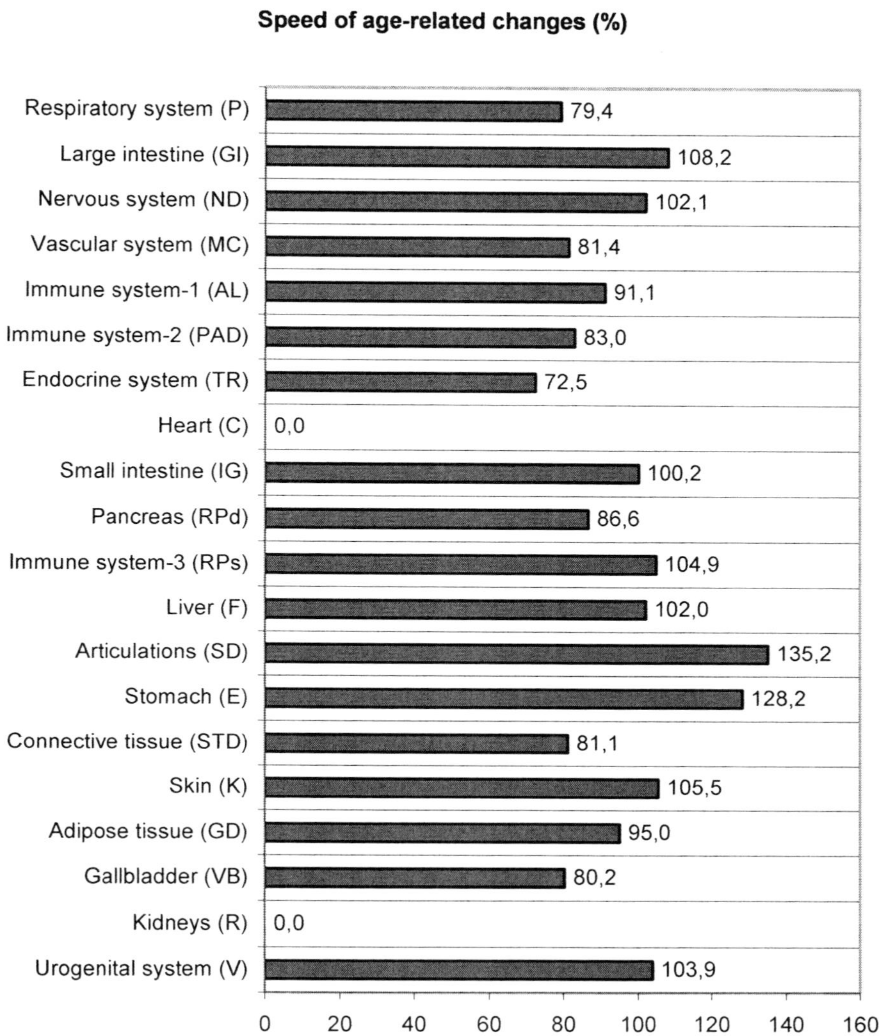

Picture 46.

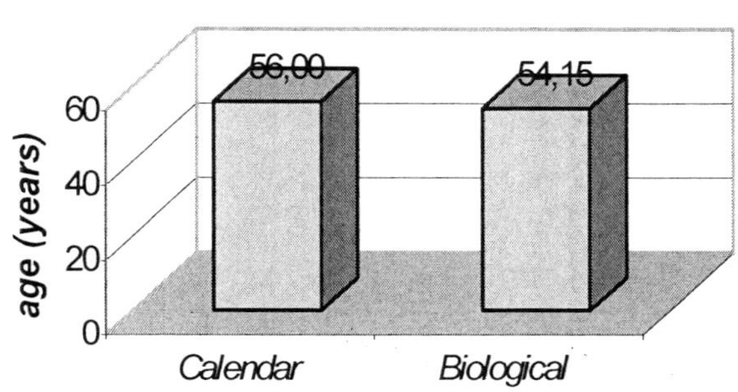

Picture 47. Comparative Analysis of biological and calendar ages.

3. Summary

(1) Reliable age-related changes have been determined by electropunctual diagnostics method;
(2) Formulas to determine BA taking into account gender, both for the whole organism and for its separate systems have been derived;
(3) Electropunctual method meets major requirements to the methods of BA determination.

Reference

[1] A.V. Samokhin, Yu.V. Gotovsky. *«Electropunctual diagnostics and R. Foll method therapy». – «Imedis»* Intellectual Medicine System Center, Moscow–1995.
[2] I.S. Rolik, A.V. Samokhin, S.E. Fursov. *"Representative electroacupuncture points reference book according to R.Foll." – MC «Systema»*, Moscow 1991.
[3] A.A. Podkolzin, V.N. Krutko, V.I. Dontsov and others. *«Quantitative evaluation of mortality, ageing, life expectancy and biological age indexes». - MGMSU*, Moscow 2001.

INDEX

A

B

C

D

E

F

G

H

I

O

P

R

S

T

U

O

P

R

V

W

Y

Z